An Introduction to Zeolite
Molecular Sieves

An Introduction to Zeolite Molecular Sieves

A. Dyer

Department of Chemistry and Applied Chemistry
University of Salford
UK

JOHN WILEY & SONS

Chichester · New York · Brisbane · Toronto · Singapore

Library of Congress Cataloging-in-Publication Data:

Dyer, A. (Alan)
  An introduction to zeolite molecular sieves.

  1. Molecular sieves. 2. Zeolites. I. Title.
TP 159.M6D84   1988   620.198   88-5524

ISBN 0 471 91981 0

British Library Cataloguing in Publication Data:

Dyer, A. (Alan)
  An introduction to Zeolite Molecular Sieves.

  1. Molecular sieves : Zeolites
  I. Title
  541.3'453

ISBN 0 471 91981 0

Typeset by Macmillan India Ltd.,
Printed and bound in Great Britain by
Bath Press Ltd., Bath, Avon

Dedication

To all my friends and colleagues in the international community of zeolite scientists and to all my coworkers who have made 30 years of zeolites such a pleasure.

Acknowledgements

I am grateful for the permission granted to reproduce information from the sources noted in the text. My thanks also to Linda Chawner for her help with various photographs and to Dr S. Ramdas (BP Sunbury) for the computer representation of zeolite Theta. In conclusion my sincere thanks to Sylvia Thomson for her excellent typing skills and help in the production of the manuscript of this book.

Contents

Introduction

To be interested in the zeolite compounds is to become involved in many aspects of science and technology. They are materials with unique properties which find uses in such diverse fields as oil cracking, nuclear waste treatment and animal feed supplementation. In their natural form they have a special importance in recent geology and their elegant molecular architecture has attracted an unrivalled interest from scientists seeking to apply modern instrumental methods to structure solving.

The current rate of publications related to zeolites and their applications is about 2000 per year and recently three or four international conferences having zeolites as their major topic have been held every year.

This book is intended to provide an introduction to zeolite science. It is hoped that it will prove useful to scientists starting out on zeolite projects and will encourage academic institutions to include in their teaching programmes such coverage as befits the scientific, technological and economic importance of the zeolite minerals. An in-depth treatment of zeolite ion exchange, syntheses and catalytic behaviour will not be attempted as excellent texts already exist to provide this knowledge.

CHAPTER 1

What is a zeolite?

Zeolites are a well-defined class of crystalline naturally occurring aluminosilicate minerals. They have three-dimensional structures arising from a framework of $[SiO_4]^{4-}$ and $[AlO_4]^{5-}$ coordination polyhedra (Fig. 1) linked by all their corners. The frameworks generally are very open and contain channels and cavities in which are located cations and water molecules (Fig. 2). The cations

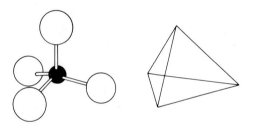

Fig. 1.   Representations of $[SiO_4]^{4-}$ or $[AlO_4]^{5-}$ tetrahedra

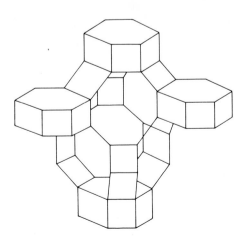

Fig. 2.   The framework structure of chabazite. Each line represents an oxygen atom and each junction a silicon or aluminium (Water molecules fill the space in the cages and cations float in this aqueous environment)

1

often have a high degree of mobility giving rise to facile ion exchange and the water molecules are readily lost and regained; this accounts for the well-known desiccant properties of zeolites.

The word 'zeolite' has Greek roots and means 'boiling stones', an allusion to the visible loss of water noted when the natural zeolites are heated. This property, of course, illustrates their easy water loss and is described as 'intumescence'.

Easy water loss has been recognized in other materials and many hydrates have been described as having 'zeolitic' water. Some textbooks describe the well-known synthetic aluminosilicates used as water softeners (Permutits) as zeolites, but this is incorrect as they are amorphous.

Many of the natural zeolites can be produced synthetically and several crystalline aluminosilicates with framework structures with no known natural counterpart have been made in the laboratory. The best known example of a synthetic zeolite is zeolite A, which can be related structurally to naturally occurring zeolites and so justify its inclusion as a zeolite mineral. It, like other synthetic zeolites, exhibits the definitive zeolitic properties of ion exchange and reversible water loss.

Another characteristic zeolite property arises from their molecular framework structures in that the assemblages of tetrahedra creating their porous structure happen to create regular arrays of apertures. These apertures are of such a size as to be able to selectively take up some molecules into their porous structure, whilst rejecting others on the basis of their larger effective molecular dimensions. This is the property of 'molecular sieving', largely unique to zeolites and responsible for their first commercial success.

It can be conjectured that every chemical laboratory in the world has on its shelves bottles of 'Molecular Sieve 3A, 4A and 5A', which are the alternative names for the ion-exchanged forms of synthetic zeolite A first produced by the Linde Corporation (now part of Union Carbide). Some carbons and silicas have molecular sieve properties so therefore some non-aluminosilicate materials have been described as 'zeolitic' because of their sieving behaviour.

Problems in definition are arising due to the ever-increasing numbers of crystalline porous substances being synthesized by zeolite-type preparative methods. These materials contain coordination polyhedra of elements other than just silicon or aluminium. A typical example is given by the '$AlPO_4$' substances, made by workers at Union Carbide, which are commonly described as zeolites.

This extension of definition is hard to justify as on this basis many diverse compounds would have to be logically described as zeolites. Many ferricyanides are molecular sieves and ion exchangers, and have framework structures. They are crystalline and show reversible water loss. Similarly hundreds of heteropoly acids and their salts (phosphates, arsenates, etc.) can be so categorized. Even organic ion exchange resins have been described as 'zeolites' and in the nuclear waste treatment literature natural and synthetic zeolite minerals are often included as 'resins'!

The only conclusion which can be sensibly reached is to reserve the word zeolite for the traditionally defined aluminosilicate minerals (natural and synthetic)— as in the first paragraph of this chapter. It is obviously unacceptable to the mineralogical world that materials *not* composed essentially of oxygen, silicon and aluminium should be described as zeolites and non-mineralogists should respect this. A new word is needed to encompass the materials which, although they do not fit the confined definition, are of relevance to zeolite science and technology. It is suggested that substances typified by the $AlPO_4$s should be called 'zeotypes' and this will be used in this text.

CHAPTER 2

Natural zeolites and their occurrence

Introduction

In 1756 a Swedish mineralogist, Crønstedt, recognized a new mineral species which he called 'zeolite' on the basis of its intumescence. He found zeolites in relatively small cavities (vugs) in rocks of volcanic origin—a classical zeolite occurrence.

In this matrix zeolites usually form as spectacular crystals (Fig. 3) and are much collected by mineralogists the world over as they are ubiquitous in nearly every basalt formation and in many rocks of similar origin. They are generally found in the more recent geological time zones but one publication cites zeolites in Cambrian rocks.

Virtually every museum has a zeolite collection and the interested reader is referred to the splendid display of zeolites from Scotland in the Royal Museum of Scotland in Edinburgh. Faujasite of Swiss and German origin can be bought in costume jewellery—particularly as earrings, cuff links and brooches.

Jewellery represented just about the only commercial interest in zeolite minerals for nearly 200 years, although Eichhorn demonstrated as early as 1858 that chabazite and natrolite exhibited reversible ion exchange. In addition Weigel and Steinhoff described molecular sieve attributes of chabazite in 1925.

These early qualitative observations were extended by the pioneering work of R.M. Barrer, which commenced in 1938 and is still continuing at the time of writing. He, above all, placed zeolite science on a firm physicochemical footing and furnished quantitative and theoretical descriptions of the ion exchange, dehydration and gas-sorptive behaviour of natural zeolites—especially chabazite. Barrer also demonstrated that some zeolites could be synthesized in a form identical to their natural counterparts and that new zeolite phases, unknown in nature, resulted from relatively simple hydrothermal reactions under laboratory conditions. At that time the emphasis largely (but not exclusively) was to mimic those conditions under which the zeolites of volcanic origin were presumed to have been formed, i.e. at high temperatures and salt concentrations and autogeneous pressures. This changed dramatically in 1949 when workers at Union Carbide, directed by R.M. Milton, synthesized zeolites by low-temperature hydrothermal processes.

(a)

(b)

Fig. 3.—Natural zeolites (a) natrolite from Co. Antrim, N. Ireland, (b) stilbite
from Skye, Scotland (Photo L.E. Chawner).

This finding triggered off the present lucrative desiccant and molecular sieve market. It had the side-effect, further stimulated in about 1960 by the discovery of zeolite catalysis, of encouraging geologists (particularly those employed by oil companies) to look for more extensive natural zeolite deposits of potential economic importance as it was not feasible to garner them from volcanic cavities in sufficient quantity to be of industrial use.

The searches have been chronicled by Mumpton and were of a quite unexpected and far-reaching success. Parts of their findings in fact existed in early literature (if well hidden!), when for instance in 1891 Murray and Renard described zeolites in 'red mud' recovered from the bed of the Pacific Ocean by HMS *Challenger* in the first ever oceanographic survey (1873–1876) and in 1914 Johannsen reported fine-grained zeolites in beds of volcanic tuffs located in Colorado, Wyoming and Utah in the USA. Other early reports noted zeolites in saline lake beds and in drill cores close to geyser activity in Yellowstone National Park, USA. These reports illustrate the now well-defined occurrences of zeolites in sedimentary and low-grade metamorphic rocks. They are dispersed worldwide and often occur in megaton quantities at high purities (> 90%). Workable deposits are located in the USA, Japan, Cuba, USSR, Italy, Czechoslovakia, Hungary, Bulgaria, South Africa, Yugoslavia, Mexico, Korea—although some of these contain only about 60% zeolite content.

In the short space of 25 years zeolites have become recognized as one of the most abundant mineral species on earth, an unparalleled change from their 'rarity' still chronicled in many museum displays. Natural zeolites still await large-scale use, but they have some commercial usage as will be described in later chapters. Current literature reflects the high interest shown in the prospects of natural zeolites as a new commodity.

The remainder of this chapter contains a more detailed description of the modes and nature of their natural occurrences.

Zeolites of saline alkaline lake origin

The creation of zeolites in a saline alkaline lacustrine environment is typified by a closed basin in arid, or semi-arid, regions in which lake water of high sodium carbonate/bicarbonate content has gathered to produce a high pH ($\sim$ 9.5). Zeolites are laid down from reactive materials deposited in the lake. These materials are often 'glasses' of volcanic origin, wind blown, and described as 'vitric tuffs'. Other substances forming zeolites under these conditions are biogenic silica (skeletons of radiolareans and forams), clays, plagioclase (feldspar) and forms of quartz.

Deposits of this type commonly contain phillipsite, clinoptilolite and erionite, but some include chabazite and mordenite. Zeolite locations of this sort are widespread in the Western USA (Fig. 4) stemming from the Plio-Pleistocene era. Older deposits (Eocene, Cenozoic) often contain analcime which arises from subsequent alteration of the zeolites in the younger rocks.

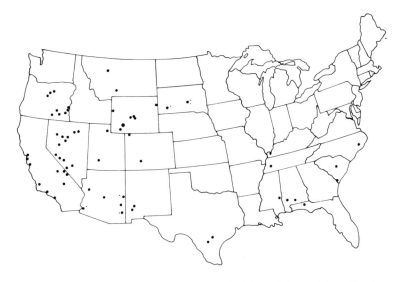

Fig. 4. Locations of clinoptilolite found in the USA and formed in saline lacustrine environments. Reprinted by permission from R. A. Sheppard in *Molecular Sieve Zeolites*—1, E. M. Flanigen and L. B. Sand (eds.), ACS Symposium Series 101; 1971, p. 300. Copyright (1971) American Chemical Society

Zeolites in soils and land surfaces

Locations of this environment are typified by the San Joaquim valley, California and the Big Sandy Formation in Arizona, USA (Fig. 5) Other well-studied sites are those in the Olduvai Gorge, Tanzania, and at Lake Magadi in Kenya. Again dry, closed basins are required. In California the 'reactive' material is mostly montmorillonite (a clay mineral) and the high pH is caused by evapotranspiration; analcime is the most abundant species. At the Olduvai Gorge wind-blown tuffs have been altered to form phillipsite, natrolite, chabazite and analcime. Geologically these deposits are young (Pleistocene and Holocene), red–brown in colour and very abundant. The zeolite content of the soils is low (15–40%) and a similar occurrence is reported in the Harwell series of soils of Berkshire (UK), which have a high heulandite content.

Zeolites in marine deposits

Marine zeolites readily form at shallow depths and low temperatures as well as at more substantial depths and higher temperatures. This latter condition will be discussed under burial diagenesis.

The *Glomar Challenger* oceanographic survey ship drilled to depths of 400–700 m in the Pacific ocean bed and found abundant occurrences of phillipsite and clinoptilolite (Fig. 6). Analcime, erionite and mordenite were also present.

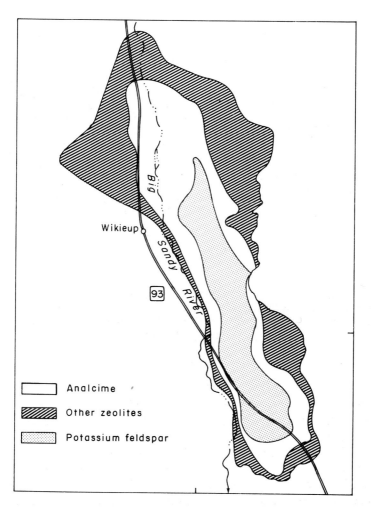

Fig. 5 Simplified map of the Big Sandy Formation in Arizona, USA, showing zeolite distributions. Reprinted with permission from R. C. Surdam and R. A. Sheppard in L. B. Sand and F. A. Mumpton (eds.), *Natural Zeolites, Occurrences, Properties, Use.* Copyright (1978) Pergamon Books Ltd

They seemed to have been formed principally from the action of trapped salt solutions (pore fluids) on glasses of underwater volcanic origin. Similar circumstances have formed zeolites under the Indian and Atlantic Oceans. In some cases biogenic silica contributes to clinoptilolite production. Analcime, present again, is thought to arise from slow alteration of other zeolite species over a period of some 100 million years. With further passage of time it is probable that further changes occur to feldspars which emerge as a decrease in zeolite content noted in rocks of marine origin from the Mezozoic and Paleozoic eras.

8

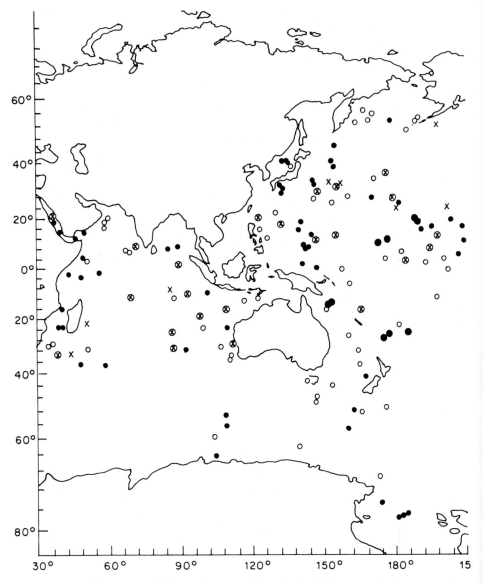

Fig. 6. Distribution of phillipsite (×) and clinoptilolite(○) in cores from ocean bed analysis. Solid points are core analyses with no zeolites present. From Boles in

9

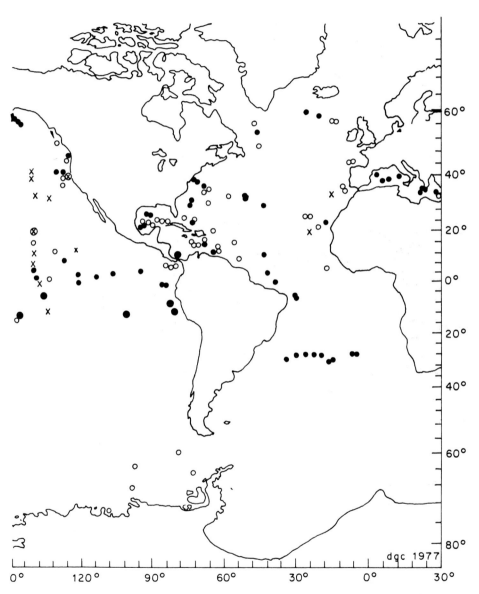

F. A. Mumpton (ed.), *Mineralogy and Geology of Natural Zeolites*, Mineralogical Society
of America, 1977

Zeolites from open flowing systems

Zeolites can be formed when flowing waters of high pH and salt content interact with vitric volcanic ash causing rapid crystal formation. Sites demonstrating this genesis have been recorded in Nevada, in the Koko Crater (Hawaii) and in the abundant tuffs in Southern Italy (typified by the yellow Neopolitan ashes). Tuffs in Southern Italy commonly contain 60% chabazite with some 10% phillipsite.

Evidence suggests that time scales as short (by geological standards) as 4000 years are needed for these formations to occur. The high pH of the system stems from hydrolysis of the volcanic ash by surface water. Clinoptilolite, analcime, phillipsite and chabazite are found in these locations.

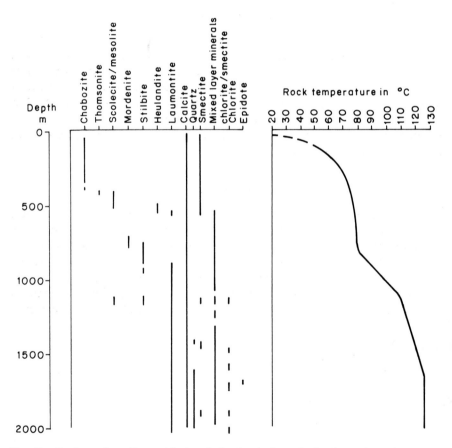

Fig. 7. Zoning of zeolites with borehole depth in a hydrothermal environment. Reprinted with permission from H. Kristmannsdotter and J. Thomasson in L. B. Sand and F. A. Mumpton (eds.), *Natural Zeolites, Occurrences, Properties, Use.* Copyright (1978) Pergamon Books Ltd

Hydrothermally treated zeolites

Zeolites are well known in Yellowstone Park (USA), in Iceland and in Wairakei (New Zealand) where they have formed from geothermal (geyser) action on existing volcanic ash. Often deposits occur in well-defined zones (Fig. 7), with clinoptilolite and mordenite forming in the shallowest and coolest zones with deeper (hotter) environments containing analcime, heulandite, laumontite and wairakite. Ferrierite, thomsonite, chabazite, mesolite, scolecite and stilbite are also known in hydrothermal zones.

An anachronism of zeolite science is worth noting here. As yet no true synthesis of laumontite has been reported (although a seeded one exists), but it has been reliably reported as being all too easily formed in the central heating system of a Russian city in Kamchatka, fed by natural hot spring water!

Zeolites formed by burial diagenesis

This classification relates to minerals formed as a result of their depth of burial, by subsequent layers of geologic species, and the consequential geothermal gradient. This group has strong associations with deep-sea and hydrothermal conditions. It reflects decreases in hydration with increased depth, so relatively porous zeolites (clinoptilolite and mordenite) are found above those of a less open nature (analcime, heulandite and laumontite).

Deposits of this nature have been described in the Green Tuff region of Japan and in the John Day formation in Oregon, USA.

Summary

It can be appreciated that zeolites are readily formed in a variety of geological environments mainly from volcanic debris. They may well be important to the formation of other minerals (e.g. feldspars and clay minerals) by alteration and similar phenomena. Zeolites have increased in their geological standing during recent years—a story still progressing and enlivened by their association with some oil-bearing rocks and speculation as to the likelihood of zeolite structural cavities being suitable environments for the generation of protein precursors.

CHAPTER 3

The structure of zeolites

Introduction

At present some 39 naturally occurring zeolite species have been recorded and their structures determined. In addition more than 100 synthetic species with no known natural counterparts have been reported in the literature. Relatively few of these have been confirmed as new zeolites and the majority await full structural determination. This chapter will describe the major structural classes of known zeolites together with a more detailed consideration of those zeolites of current commercial significance.

Fundamental zeolite structural units

As stated earlier all zeolites have framework (three-dimensional) structures constructed by joining together $[SiO_4]^{4-}$ and $[AlO_4]^{5-}$ coordination polyhedra. By definition these tetrahedra are assembled together such that the oxygen at each tetrahedral corner is shared with that in an identical tetrahedron (Si *or* Al), as shown in Fig. 8. This corner (or vertex) sharing creates infinite lattices comprised of identical building blocks (unit cells) in a manner common to all crystalline materials.

One way to classify zeolite structures would be to relate them to the symmetry of their unit cells. This would be cumbersome and is much simplified by the

Fig. 8.   Tetrahedra linked together to create a three-dimensional structure

observation that zeolite structures often have identical (or very similar) repeating structural sub-units which are less complex than their repeating unit cells.

These recurring units are called 'secondary building units' (sbus) and the simplest, most utilitarian, classification describes all known zeolite frameworks as arrangements linking eight sbus shown in Fig. 9. These denote only the aluminosilicate skeleton (i.e. the Si, Al and O positions in space relative to each other) and exclude consideration of the cation and water moieties sited within the cavities and channels of the framework. The cation and water sites are complex and only fully defined in certain zeolites as will become apparent later.

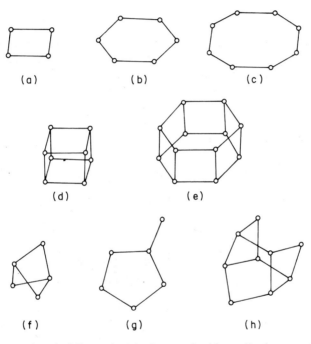

Fig. 9   The secondary building units (sbus) recognized in zeolite frameworks; (a) single four ring (S4R), (b) single six ring (S6R), (c) single eight ring (S8R), (d) double four ring (D4R), (e) double six ring (D6R), (f) complex 4-1, (g) complex 5-1 and (h) complex 4-4-1

For the present it should be noted that the number of cations present within a zeolite structure is determined by the number of $[AlO_4]^{5-}$ tetrahedra included in the framework. This arises from the isomorphous substitution of Al^{3+} for Si^{4+} into the component polyhedra, causing a residual negative charge on the oxygen framework. This negative charge is compensated by those cations present in the synthesis and held in the interstices of the structure on crystallization. The extent and location of water molecule incorporation depends upon (i) the overall architecture of the zeolite molecular structure, i.e. the size and shape of the cavities and channels present, and (ii) the number and nature of the cations in the structure.

14

The aluminosilicate skeleton can be represented in a number of ways, as for example in the traditional 'ball and stick' model (Fig. 10). The most favoured is the use of tetrahedral arrays, adopted by organic chemists, where the oxygen atoms are drawn as single 'bonds' joining together tetrahedral 'centres' depicting silicon and aluminium. This is the method used in Fig. 9. Further perusal of Fig. 9 shows that each sbu contains rings of tetrahedra which are equivalent to rings of oxygen atoms described as 'single four rings', 'single six rings' etc. (Fig. 11). When

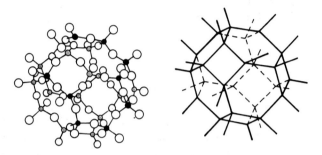

Fig. 10   A 'ball and stick' representation of the structure of the sodalite unit (left) with a framework diagram (like Fig. 2) for comparison

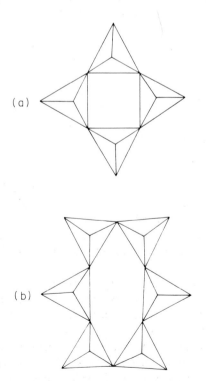

Fig. 11.   The arrangement of tetrahedra in the (a) S4R and (b) S6R sbu

the sbus are joined to create the infinite lattices they can proscribe larger rings containing 8, 10 or 12 linked tetrahedra (i.e. rings of 8, 10 or 12 oxygen atoms). These rings are obviously important structural features and are often called 'oxygen windows'.

Classification of zeolite structure by sbu content

Tables I–V list known zeolite structures classified by (i) their sbu content, (ii) structure type (IUPAC nomenclature), (iii) name (prefix 'zeolite' means that the material is known only as a synthetic material) and (iv) typical unit cell content.

Comments on Tables I–V

As with all classifications there are queries and possible anomalies in the lists of structures given in these tables. It can be seen (Table V) that barrerite is the sodium form of stellerite and that they have the same framework. There are other equally obvious examples of the same resemblance (see the natrolites in Table III). The justification for their inclusion rather than the quotation of the frameworks alone is that the list includes all the known zeolite *minerals* recorded and confirmed in the literature. Omitted from the tables are (a) some synthetic zeolites unlikely to be of commercial interest, (b) 'new zeolites' whose structures have not yet been determined, (c) zeolite-like minerals and (d) zeotypes. Table VI lists examples from these categories with comments.

Other points to note are the following:
(1) Zeolite Y is isostructural with zeolite X (and faujasite). It differs in its higher Si:Al ratio (in the relatively lower amount of isomorphous substitution of Al for Si into tetrahedral framework position) and this causes a drop in the number of cations present in the framework and an increase in the number of water molecules present.
(2) Zeolite A also has a structural isotype described as ZK-4.
(3) Mazzite is the first recorded example of a zeolite known initially as a synthetic material (zeolite Omega) and later found in nature.
(4) Mordenite is described in various publications as 'large and small port'. This distinction relates to two types of mordenite having different molecular sieve properties. There is no firm evidence for any framework structural variation between these two varieties.
(5) Varieties of sodalite and cancrinite have been included despite the fact that, mineralogically speaking, they are not zeolites but felspathoids. They are in the tables because their cage structures (Fig. 12) play important roles in some zeolite frameworks. When synthesized with the unit cell compositions quoted they possess zeolitic properties. In nature they contain salt molecules trapped inside their cages—hence the mineralogical distinction. The salts can be washed out leaving materials of zeolitic type. Sodalite hydrate often is described as 'hydroxy sodalite' (HS).

Table I. Zeolite structures based upon single oxygen ring sbus

Secondary building unit (sbu)	Structure type	Name	Typical unit cell content
S4R	ANA	Analcime	$Na_{16}Al_{16}Si_{32}O_{96}10H_2O$
	ANA	Wairakite	$Ca_8Al_{16}Si_{32}O_{96}16H_2O$
	GIS	Gismondine	$Ca_4Al_8Si_8O_{32}16H_2O$
	GIS	Amicite	$K_4Na_4Al_8Si_8O_{32}10H_2O$
	GIS	Garronite	$NaCa_{2.5}Al_6Si_{10}O_{32}14H_2O$
	GIS	Gobbinsite	$Na_5Al_5Si_{11}O_{32}11H_2O$
	GIS	Zeolite NaP-1	$Na_6Al_6Si_{10}O_{32}12H_2O$
	LAU	Laumontite	$Ca_4Al_8Si_{16}O_{46}16H_2O$
	MER	Merlionite	$K_5Ca_2Al_9Si_{23}O_{64}24H_2O$
	PAU	Paulingite	$(K_2,Na_2,Ca,Ba)_{76}Al_{152}Si_{520}O_{1344}700H_2O$
	PHI	Phillipsite	$K_2Ca_{1.5}NaAl_6Si_{10}O_{32}12H_2O$
	PHI	Harmotome	$Ba_2Ca_{0.5}Al_5Si_{11}O_{32}12H_2O$
	YUG	Yugawaralite	$Ca_2Al_4Si_{12}O_{32}8H_2O$
S6R	CAN	Cancrinite hydrate	$Na_6Al_6Si_6O_{24}8H_2O$
	ERI	Erionite	$Na_2K_2Mg_{0.5}Ca_2Al_9Si_{27}O_{72}27H_2O$
	LEV	Levynite (Levyne)	$NaCa_3Al_7Si_{11}O_{36}18H_2O$
	LTL	Zeolite L	$K_6Na_3Al_9Si_{27}O_{72}21H_2O$
	LOS	Zeolite Losod	$Na_{12}Al_{12}Si_{12}O_{48}19H_2O$
	MAZ	Mazzite (Zeolite Omega)	$Mg_2K_3Ca_{1.5}Al_{10}Si_{26}O_{72}28H_2O$
	OFF	Offretite	$KCa_2Al_5Si_{13}O_{36}15H_2O$
	SOD	Sodalite hydrate (HS)	$Na_6Al_6Si_6O_{24}8H_2O$
S8R			Occurs in many structures but with other sbus (see structure of zeolite A, chabazite, etc.)

Table II. Zeolite structures based upon double oxygen ring sbus

Secondary building unit (sbu)	Structure type	Name	Typical unit cell content
D4R	LTA	Zeolite A	$Na_{12}Al_{12}Si_{12}O_{48}27\,H_2O$
D6R	CHA	Chabazite	$Ca_2Al_4Si_8O_{24}13\,H_2O$
	CHA	Wilhendersonite	$K_2Ca_2Al_6Si_6O_{24}10\,H_2O$
	FAU	Faujasite	$Na_{12}Ca_{12}Mg_{11}Al_{58}Si_{134}O_{384}235\,H_2O$
	FAU	Zeolite X	$Na_{88}Al_{88}Si_{104}O_{384}220\,H_2O$
	GME	Gmelinite	$Na_8Al_8Si_{16}O_{48}24\,H_2O$
	KFI	Zeolite ZK-5	$Na_{30}Al_{30}Si_{66}O_{192}98\,H_2O$
	RHO	Zeolite Rho	$(Na,Cs)_{12}Al_{12}Si_{36}O_{96}46\,H_2O$

Table III. Zeolite structures based upon the 4–1 sbu

Secondary building unit (sbu)	Structure type	Name	Typical unit cell content
4–1	EDI	Edingtonite	$Ba_2Al_4Si_6O_{20}8\,H_2O$
	NAT	Natrolite	$Na_{16}Al_{16}Si_{24}O_{80}16\,H_2O$
	NAT	Tetranatrolite	$Na_{16}Al_{16}Si_{24}O_{80}16\,H_2O$
	NAT	Paranatrolite	$Na_{16}Al_{16}Si_{24}O_{80}24\,H_2O$
	NAT	Mesolite	$Na_{16}Ca_{16}Al_{48}Si_{72}O_{240}64\,H_2O$
	NAT	Scolecite	$Ca_8Al_{16}Si_{24}O_{80}24\,H_2O$
	THO	Thomsonite	$Na_4Ca_8Al_{20}Si_{20}O_{80}24\,H_2O$
	THO	Gonnardite	$Na_5Ca_2Al_9Si_{11}O_{40}14\,H_2O$

Table IV. Zeolite structures based upon the 5–1 sbu

Secondary building unit (sbu)	Structure type	Name	Typical unit cell content
5-1	BIK	Bikitaite	$Li_2Al_2Si_4O_{12}2H_2O$
	DAC	Dachiardite	$Na_5Al_5Si_{19}O_{48}12H_2O$
	EPI	Epistilbite	$Ca_3Al_6Si_{18}O_{48}16H_2O$
	FER	Ferrierite	$NaCa_{0.5}Mg_2Al_6Si_{30}O_{72}20H_2O$
	MFI	Zeolite ZSM-5	$Na_nAl_nSi_{96-n}O_{192}\sim16H_2O(n\sim3)$
	MOR	Mordenite	$Na_8Al_8Si_{40}O_{96}24H_2O$

Table V. Zeolite structures based upon the 4-4-1 sbu

Secondary building unit (sbu)	Structure type	Name	Typical unit cell content
4-4-1	BRE	Brewsterite	$Sr_2Al_4Si_{12}O_{32}10H_2O$
	HEU	Heulandite	$Ca_4Al_8Si_{28}O_{72}24H_2O$
	HEU	Clinoptilolite	$Na_6Al_6Si_{30}O_{72}24H_2O$
	STI	Stilbite	$Na_2Ca_4Al_{10}Si_{26}O_{72}34H_2O$
	STI	Stellerite	$Ca_4Al_8Si_{28}O_{72}28H_2O$
	STI	Barrerite	$Na_8Al_8Si_{28}O_{72}26H_2O$

19

Fig. 12. The cancrinite (left) and sodalite cages

(6) Two variations in notation commonly arise. Analcime is often referred to as 'analcite' and levynite as 'levyne'. Early literature had many synonyms for the natural zeolites and some of the more common examples are listed in Table VI.

Table VI
(a) Other natural and species cited in zeolite literature

Natural species	Comment
Afghanite(AFG) Liottite(LIO) Franzinite Sacrofanite Giuseppettite	All variations on cancrinite or sodalite structures
Svetlozarite	Multiply twinned and highly faulted dachiardite
Doranite	Analcime with a high magnesium content
Chiarennite	A beryllium silicate similar to bikitaite (zeotype)
Hsianghualite	A beryllium silicate (another zeotype)
Lovdarite(LOV)	$Li_{16}Ca_{24}Be_{24}Si_{24}O_{96}F_2$ (zeotype)
Wenkite(WEN)	$Na_{0.5}K_{0.5}Ca_{5.5}Ba_{3.5}Si_{11}Al_9O_{41}(OH_2)(SO_4)_3 H_2O$
Roggianite(ROG)	$Ca_{16}(Al_{16}Si_{32}O_{88}OH_{16})[OH]_{16} 26 H_2O$
Partheite(PAR)	$Ca_8Al_{16}Si_{16}O_{64}16 H_2O$
Perlialite	May be isostructural with Zeolite L
Viseite	$Na_2Ca_{10}(Al_{20}Si_6P_{10}O_{60}(OH)_{36})16 H_2O$ (a natural zeotype with analcime framework)
Keoheite	Naturally occurring phosphoaluminate with the analcime framework (another natural zeotype)
Leucite	Felspathoid with analcime framework ($K_{16}Al_{16}Si_{32}O_{96}$)
Pollucite	Caesium felspathoid with analcime framework ($Cs_{16}Al_{16}Si_{32}O_{96}$)
Herschelite, phacolite	Chabazites with pseudo-hexagonal crystal habits
Leonhardite	Partially dehydrated laumontite
Wellsite	Barium variety of phillipsite
Goosecreekite(GOO)	$Ca_6Al_2Si_{12}O_{32}10H_2O$ Structure of 4-, 6- and 8- membered rings with some resemblances to brewsterite but not allocated as yet to the groupings in Table IV
Cowlesite	$Ca_6Al_2Si_{18}O_{60}36 H_2O$ (structure unknown)

Table VI (contd.)

(b) Outmoded names for zeolites

Natural species	Old names
Chabazite	Acadialite, haydenite
Epistilbite	Parastilbite, orizite, reissite
Gismondine	Zeagonite
Laumontite	Caporcianite, lomontite, scheiderite, sloanite
Mordenite	Ptilolite, arduinite, pseudo natrolite
Natrolite	Galaktite
Phillipsite	Christianite
Stellerite	Epidesmine
Stilbite	Desmine, foresite
Thomsonite	Fareolite, comptonite

(c) Synthetic zeolite phases[a]

Name	Comment
Li-A(BW)(ABW)	Structure known
TMA-E(AB)(EAB)	Structure known
Linde type N(LTN)	Structure known
Na,K-F	Isostructural with zeolite F,Rb-D, edingtonite
P[b],Q	ZK-5 structures
Nu, Fu	Structure known
Theta-1 (TON)	The only 'two-dimensional zeolite' structure known
O	Offretite phase
B	Gismondine structure
K-M,W	Merlionite structure
Ba-G	Isotype of zeolite L
Pentasils	Generic names for the ZSM-5 (MFI), ZSM-11 (MEL) family of structures
Dodecasils	Phases of silica[c] with some significant zeolite properties
ZSM-39 (MTN)	(also called 'clathrosils' because of their structural simil-
Silicalite	arities to clathrate hydrates)
Nonasil (NON)	
ZSM-23 (MTT)	Structure known
ZSM-12 (MTW)	Structure known
ZSM-3	Polymorph of faujasite
ZSM-4	Mazzite structure
ZSM-21	Ferrierite structure
ZSM-34	Offerite/erionite structure intergrown

[a] Not a comprehensive list, excludes many species not yet defined.
[b] Not to be confused with Na-P.
[c] Melanophlogite (MEP) is a naturally occurring phase of silica with zeolite properties (not yet synthesized).

Comments on Table VI

(1) The atlas of known zeolite structures (see Bibliography) also lists structure types associated with silica phases and related materials, e.g. NON (nonasil), DOH (dodecasil-1H) and DDR (deca-dodecasil-3R).

(2) Lovdarite, wenkite and roggianite have been included in the most recent atlas of structures.

Specific zeolite topologies

Introduction

The International Zeolite Association (IZA) has published a handbook which presents a further, more detailed, view of zeolite frameworks. The handbook describes some 40 currently known zeolite topologies and assigns all known

Table VII. Zeolite frameworks recognised by the Structure Commission of the IZA[a]

Structure	Code	Structure	Code
Li-A(BW)	ABW	Linde type A	LTA
Afghanite	AFG	Linde type L	LTL
Analcime	ANA	Linde type N	LTN
Bikitaite	BIK	Mazzite	MAZ
Brewsterite	BRE	ZSM-11	MEL
Cancrinite	CAN	Melanophlogite	MEP
Chabazite	CHA	Merlionite	MER
Dachiardite	DAC	ZSM-5	MFI
Deca-dodecasil-3R	DDR	Mordenite	MOR
Dodecasil-1H	DOH	ZSM-39	MTN
TMA-E(AB)	EAB	ZSM-23	MTT
Edingtonite	EDI	ZSM-12	MTW
Epistilbite	EPI	Natrolite	NAT
Erionite	ERI	Nonasil	NON
Faujasite	FAU	Offretite	OFF
Ferrierite	FER	Partheite	PAR
Gismondine	GIS	Paulingite	PAU
Gmelinite	GME	Phillipsite	PHI
Goosecreekite	GOO	Rho	RHO
Heulandite	HEU	Roggianite	ROG
ZK-5	KFI	Sodalite	SOD
Laumontite	LAU	Stilbite	STI
Levyne	LEV	Thomsonite	THO
Liottite	LIO	Theta-1	TON
Losod	LOS	Wenkite	WEN
Lovdarite	LOV	Yugawaralite	YUG

[a] Not all these structures have yet been allocated to the formal groupings based upon sbu as listed in Table I.

zeolites into these topological categories. Each framework has a code (see Tables I–V) and a full list is given in Table VII.

It is beyond the scope of this text to illustrate each topology, but some zeolite structures will be described in more detail because of their scientific significance. This will provide an additional opportunity to show how structural variations occur in zeolite families in a way which cannot be appreciated very easily from sbu classifications.

Each example chosen will include descriptions of the sites of cations and water molecules within the frameworks. A complete record of all known sitings can be found in the IZA compilation (see Bibliography). It must be stressed that cation and water sitings in *any* zeolite framework are a function of temperature, water content, cation type and Si:Al ratio. In discussing this structural detail it has been decided to use the Angstrom (Å) unit rather than the correct SI unit (pm) to describe dimensional parameters. This decision has been taken in recognition of the fact that most of the zeolite literature (past and present) still uses Å rather than pm.

Examples are listed below and it should be noted that 'family' links do not always follow the IZA assignments as is evident from the first pair discussed, i.e. chabazite (CHA) and gmelinite (GME).

Chabazite and gmelinite

The framework of chabazite is a sequence of D6R units joined via S4Rs into a 'layer' pattern of the same sequence as that in cubic close packing, i.e. ABC ABC ABC. This can be described as a NaCl-type arrangement with a D6R unit at each corner of a simple cubic assemblage. Fig. 13 illustrates the ABC sequence and Fig. 14 has the same structure rotated through 45° to show the 'NaCl' array.

At the centre of this element of structure is a large cavity the entry to which is governed by passage through an aperture of eight oxygens (i.e. an eight-oxygen window or an S8R). Eleven specific cation and water locations have been identified. In the natural mineral some calcium ions are inside the D6R in the dehydrated state, but in the fully hydrated mineral these cations have a complete coordination sphere of water as they 'float' inside the large cavity.

Gmelinite has a structure based upon an hexagonal close-packing array of D6Rs in that they are linked via S4Rs into an AB AB AB repeating layer sequence (Fig. 15). Cation sites for four calcium ions have been shown adjacent to the D6Rs, when hydrated they are in the large cavities.

Mordenite (MOR) and ferrierite (FER)

In mordenite the 5-1 units are linked into a series of chains joined together to form two major channels, one restricted by 12 oxygen windows and one by eight oxygen windows. Eight cation and water sites have been identified and the major ones are (i) in the centres of the S8Rs comprising one channel, (ii) adjacent to the

Fig. 13. Chabazite structure illustrating the ABC sequence of constituent D6R sbu (Photograph L. E. Chawner.)

five oxygen rings linking the larger 12 'O' ring channels and (iii) coordinated to the S4Rs in adjacent rings (Fig. 16).

The ferrierite structure (Fig. 17) resembles that of mordenite but the chains come together to create a series of channels bound by six, eight and ten oxygens, respectively. The major (10'0') channel runs along the c crystallographic axis and cavities occur at intersections with the other channels. Seven cation/water sites have been suggested. Of these A, C and D are located in cavities occupied by two magnesium ions surrounded by four water molecules in the natural zeolite, B is occupied by sodium and lies close to four oxygens in the major channels, G and F are in S6Rs and S8Rs at the centre of the ring.

ZSM-5 (MFI) and ZSM-11 (MEL)

These zeolites can be regarded as end-members of the pentasil family. They both have structures generated by the stacking of the layers shown in Fig. 18. In

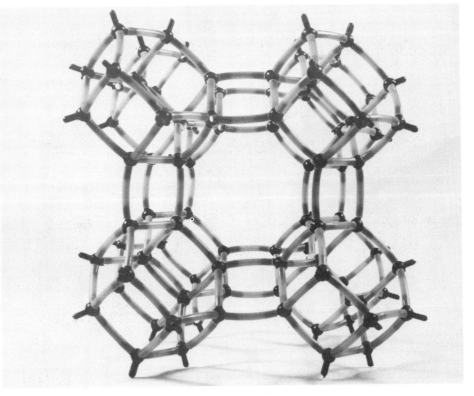

Fig. 14. Chabazite viewed to show its NaCl type array. (Photograph L. E. Chawner.)

ZSM-11 neighbouring layers are related by reflection whilst in ZSM-5 the relationship between layers is by inversion. The other family members have intermediate structures or intergrowths of ZSM-5 and ZSM-11. Both of these synthetic end-members have three-dimensional pore structures enclosed by 10 oxygen windows but these do not link cavities as such (Fig. 19).

In ZSM-5 there are two pore systems, one consisting of zig-zag channels of near-circular cross-section and another of straight channels of elliptical shape. All the intersections in ZSM-5 are of the same size.

In ZSM-11 all channels are of elliptical cross-section which results in two types of intersection, one having free space similar to that in ZSM-5 and the other being larger by about 20%.

These channels have important consequences in the zeolite sorptive and catalytic properties and even these relatively minor structural differences between ZSM-5 and ZSM-11 result in significant changes in catalytic behaviour.

These zeolites have a low aluminium content and consequently few cations are present in, them as yet unresolved, crystallographic sites. Similarly their water contents are low—indeed their frameworks have hydrophobic tendencies, in

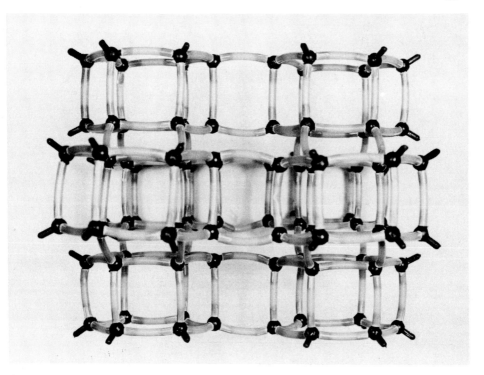

Fig. 15. Gmelinite structure illustrating the AB AB AB sequence of constituent D6R sbus. (Photograph L. E. Chawner.)

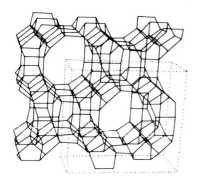 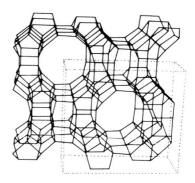

Fig. 16. The structure of mordenite as a stereopair. Reproduced with permission from W. M. Meier and D. H. Olson: *Atlas of Zeolite Structure Types* (1978). IZA Structure Commission

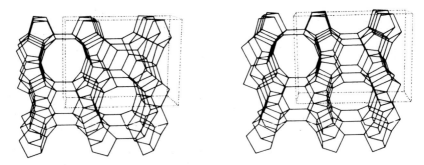

Fig. 17. The structure of ferrierite as a stereopair. Reproduced with permission from W. M. Meier and D. H. Olson: *Atlas of Zeolite Structure Types* (1978). IZA Structure Commission

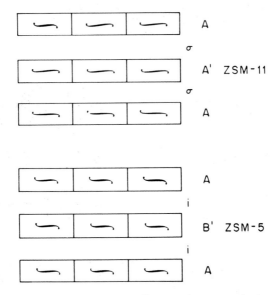

Fig. 18. ZSM-11 and ZSM-5 structures illustrated as a stacked sequence of layers. ⌒ represents a screw axis running into the plane of the paper down a series of linked 5-1 sbus forming a chain. σ describes the reflection relationship between layers A and A' and *i* describes the inversion relationship between A and B'

sharp contrast to the normally highly hydrophilic character of most other known zeolites.

Clinoptilolite and heulandite

Both zeolites are assigned to the same framework (HEU) in which 4-4-1 units are joined in a layer-like array. The layers are joined to create eight- and ten-

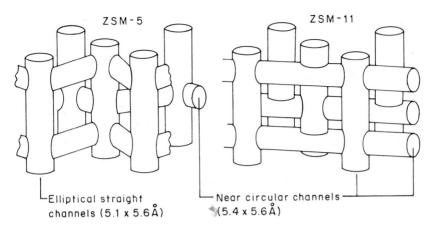

Fig. 19   Channel arrangements in ZSM-5 and ZSM-11. Reproduced by permission of the Society of Chemical Industry from A. Dyer and J. Dwyer, *Chemistry and Industry*, April, 1984 (No. 7), p. 240

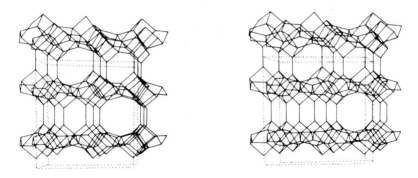

Fig. 20   The clinoptilolite/heulandite structure as a stereopair. Reproduced with permission from W. M. Meier and D. H. Olson: *Atlas of Zeolite Structure Types* (1978), IZA Structure Commission

membered oxygen windows (Fig. 20) which circumscribe two channels (both 10'0') parallel to the *c* crystallographic axis and one channel (8'0') parallel to the *a* axis.

The differences between heulandite and clinoptilolites can be based upon Si:Al ratio (heulandite <4, clinoptilolite >4), but a more reliable distinction is probably that based upon cation content. Clinoptilolites are alkali metal rich (Na + K > Ca + Mg), whereas heulandites are alkaline earth metal rich (Ca + Sr + Ba > Na + K). In clinoptilolites four cation sites have been found in the hydrated mineral, two (M1 and M2) are in the main channels, one (M3) is in the 8'0' channels and M4 is on a site close to the centre of an S8R.

28

Phillipsite (PHI) and gismondine (GIS)

In phillipsite the S4R rings are linked to form chains resembling a double crankshaft (Fig. 21). When these chains are assembled into the structure, they create S8Rs between the chains. The cations reside close to these S8Rs. The orientation of the four tetrahedra in the phillipsite chains is such that two adjacent tetrahedra point up and two point down (UUDD). Gismondine (GIS) has a related structure but with the chains linked in a different manner (Fig. 22).

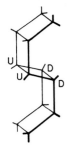

Fig. 21   The 'double crankshaft' chain element of structure in phillipsite, gismondine and related zeolites. (Note: U = up, D = down, describes the configuration of the linked tetrahedra.)

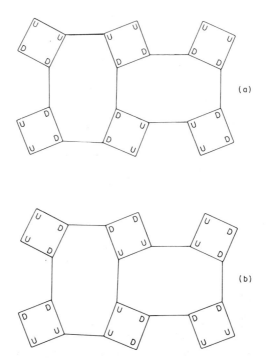

(a)

(b)

Fig. 22.   Shorthand notation of (a) phillipsite and (b) gismondine chain orientations (U = up, D = down)

These ladders of four rings can generate cancrinite, offretite and gmelinite frameworks but most classifications, as herein, prefer to use the D6R or S6R sbus to describe these zeolites.

Analcime (ANA)

Analcime has a closely arranged structure of S4Rs linked to form both S6Rs and S8Rs. The result is an unusual structure comprised of three distinct channels (two circumscribed by eight oxygens and one by six oxygens) running through the unit cell (Fig. 23). Cavities are created containing cation and water molecules (Fig. 24).

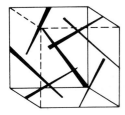

Fig. 23.   The channels in the analcime structure

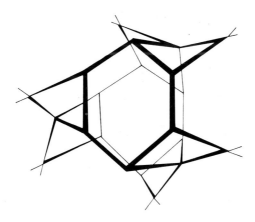

Fig. 24.   The analcime cavity viewed through an S6R

Offretite (OFF) and Zeolite L

The easiest way to describe the offretite structure is as a sequence of layers joined via both S4Rs and S6Rs to produce large channels contained by 12 oxygens (Fig. 25). Zeolite L has a similar structure (LTL) (Fig. 26). Offretite has six cation sites, one at the centre of a D6R, two associated with small cages adjacent to the main channels and three close to S6Rs.

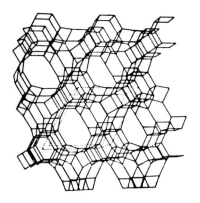 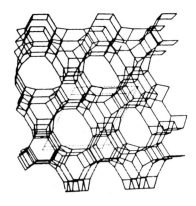

Fig. 25. The structure of offretite as a stereopair. Reproduced with permission from W. M. Meier and D. H. Olson: *Atlas of Zeolite Structure Types* (1978). IZA Structure Commission

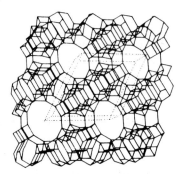 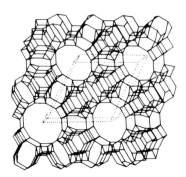

Fig. 26. The structure of zeolite L as a stereopair. Reproduced with permission from W. M. Meier and D. H. Olson: *Atlas of Zeolite Structure Types* (1978). IZA Structure Commission

Table I shows that sodalite (SOD) and cancrinite (CAN) have strong structural relationships with offretite and zeolite L but it is more relevant to this section to describe the cage structure created in SOD and CAN topologies (see Fig. 12) as this occurs in other zeolites and plays a significant role in their properties.

Natrolite (NAT)

In natrolite 4-1 units link to form a chain rather like the phillipsite structure. The '4' unit (i.e. four tetrahedra) contains the orientation UDUD and successive UDUD units join via the '1' unit to create the natrolite chains (Fig. 27).

In natrolite the linkage causes a channel and cavity structure accessed by distorted S8Rs and the cations sit in sites close to four oxygens, two each from

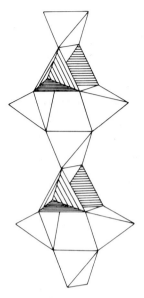

Fig. 27. The U and D arrangement of chains in the natrolite structure. (Shaded tetrahedra point up, open tetrahedra point down.) Reproduced by permission of Academic Press from R. M. Barrer, *Zeolites and Clay Minerals as Sorbents and Molecular Sieves*, 1978

adjacent 4-1 chains. Other related family members (e.g. mesolite and scolecite) have structures derived from different orientational links between 4-1 chains.

Zeolites A(LTA), X and Y (FAU)

Although these zeolites are classified in Table II on the basis of their component sbus this is not particularly appropriate to a more detailed understanding of their useful properties. An alternative view is to consider them as being composed of polyhedral cavities. In zeolite A these polyhedra are a cube synonymous with the D4R sbu, a truncated octahedron identical to the sodalite cage and a large cavity with 48 sides (Fig. 28).

The sodalite cage is often described as the β cage and the larger cavity, which is a Type I 26-hedron or truncated cubo-octahedron, is called the α cage. When these three polyhedra come together to create the A framework they each share all their faces with other polyhedra. This is called a 'space-filling' structure (Fig. 29).

Yet another way to consider the A structure is to imagine that each sodalite unit (β cage) is linked via every S4R to propagate cubic symmetry. In this cubic arrangement each sodalite cage occupies an analogous position to the atoms in a NaCl structure (see the chabazite structure in Fig. 14).

In the FAU framework, shared by the synthetic zeolites X and Y as well as the natural faujasite, the composite polyhedra are the hexagonal prism (i.e. a D6R), the sodalite cage and the Type II 26-hedron (Fig. 30). Again in the FAU framework the polyhedra are linked in a 'space filling' way (Fig. 31).

Another view would be to be aware that the sodalite units are now linked by one half of their available S6R units to produce another framework of cubic

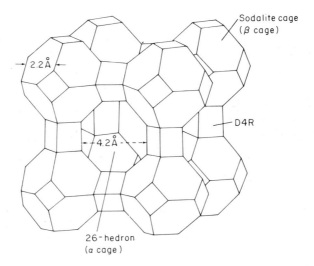

Fig. 28.   The polyhedral units comprising the structure of zeolite A

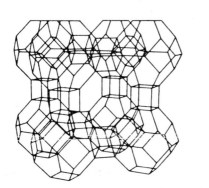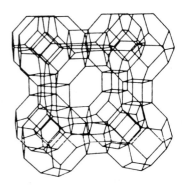

Fig. 29.   The structure of zeolite A as a stereopair. Reproduced with permission from W. M. Meier and D. H. Olson: *Atlas of Zeolite Structure Types* (1978). IZA Structure Commission

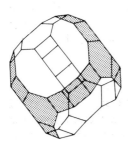

Fig. 30.   The Type II 26-hedron cavity of the faujasite structure

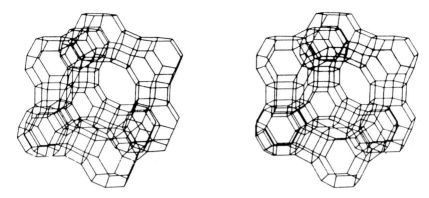

Fig. 31. The structure of faujasite as a stereopair (also that of the synthetic zeolites X and Y). Reproduced with permission from W. M. Meier and D. H. Olson: *Atlas of Zeolite Structure Types* (1978). IZA Structure Commission

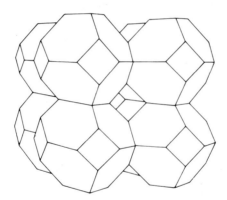

Fig. 32. The sodalite structure illustrating the total face sharing of each cage to give a space-filling structure

symmetry in which the sodalite cages occupy positions analogous to adjacent C atoms in a diamond lattice. In sodalite itself the cages pack in a 'space-filling' way with each other (Fig. 32) as do the cages in cancrinite.

This rather involved description of zeolite frameworks is very illustrative of the approach to solid-state structures in general, i.e. there always is more than one way of describing the way atoms or ions assemble into solid array and the choice of the most appropriate description should be that which is the easiest to comprehend or illustrate. Hopefully this will then coincide with the most descriptive of the properties of the materials under discussion.

The cation site occupancy in zeolites X and Y will not be dealt with in detail here, but generally it depends upon the framework charge, the nature of the cation and the amount of water present. Broadly an increase in Si:Al ratio (decrease in framework charge) depopulates the sites I and I′ in the hydrated zeolites, i.e. the cations present prefer a water environment to the framework coordination positions. The absence of water causes an increase in the population of sites I and I′ regardless of framework charge. The major sites in the faujasite framework are shown diagrammatically in Fig. 33.

When considering zeolite A it is common practice to refer to the element of structure shown in Fig. 29. This is not the full unit cell and is often described as the 'pseudo-unit cell'. The full unit cell contains 12 sodalite (β) cages. Again cation sites depend upon water content and cation type, but in the hydrated sodium form of A, of the 12 sodium ions per pseudo-unit cell, eight occupy Site I adjacent to the S6R openings to the β cages, three occupy Site II near the centre of the S8R openings to the large (α) cage and one is sited in the centre of the large

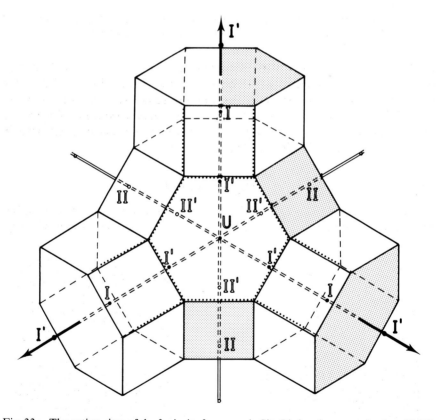

Fig. 33. The cation sites of the faujasite framework. Site I is in a hexagonal prism (D6R), I′ is near the entrance to a hexagonal prism in the sodalite (β) cage, II is inside the sodalite cage near the S6R entrances to the large (α) cage, II is in the large cage adjacent to S6R and U is at the centre of the sodalite cage. Other sites (e. g. IV and V) are in the large cavities. Reprinted with permission from J. V. Smith in *Molecular Sieve Zeolites*-1, E. M. Flanigen and L. B. Sand (eds.), ACS Symposium Series, 101; 1971, p. 173. Copyright (1971) American Chemical Society

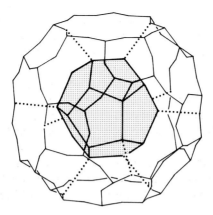

Fig. 34.   The water clathrate in the α cage of hydrated zeolite A. It is composed of 20 H₂O molecules arranged in a pentagonal dodecahedron

cavity (α cage). Each pseudo-unit cell contains 27 water molecules and 20 of these form a pentagonal dodecahedron which 'lines' the α cage (Fig. 34). This is the same arrangement of water molecules noted in some water clathrates—a fact that might well be important to the mechanism of zeolite synthesis. The residue of water molecules are in the β cages (~ 4) or held loosely inside the clathrate in association with the Na^+ ions at the centre of the α cage.

Other zeolites usefully described via cage assemblages

The synthetic materials ZK-5 (KFI) and zeolite Rho (RHO) are further examples of structures best described by their component cages as shown in Figs.

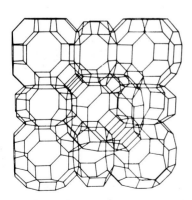
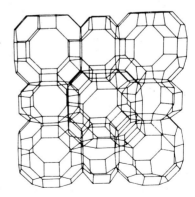

Fig. 35.   The structure of zeolite ZK-5 as a stereopair, illustrating the space-filling arrangements of the component cages (cf. Fig. 32). Reproduced with permission from W. M. Meier and D. H. Olson: *Atlas of Zeolite Structure Types* (1978). IZA Structure Commission

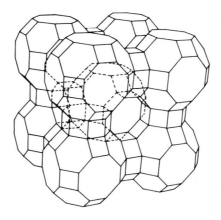

Fig. 36. The structure of zeolite Rho, illustrating the space-filling nature of its structure (cf. Figs. 32 and 35). Reprinted with permission from H. E. Robson *et al.* in *Molecular Sieves*, W. M. Meier and J. B. Uytterhoeven (eds.), ACS Symposium Series 121; 1973, p. 108. Copyright (1973) American Chemical Society

Table VIII. Some dimensional parameters of common zeolites

Zeolite	Restricting windows (number of O atoms)	Effective window size[a] (Å)	Void volume[b]	Other dimensions
Zeolite A	6	2.3	0.47	β cage diameter: 6.6 Å
	8	4.5		α cage diameter: 11.4 Å
Zeolite X	6	2.3	0.53	α cage diameter: 11.8 Å
	12	7.8		
Zeolite L	12	7.1	0.28	
Heulandite	8	4.0 × 5.5	0.35 ⎫ Interconnected	
	10	4.4 × 7.2	⎭	
Phillipsite	8	4.2 × 4.4	0.30 ⎫	
	8	2.8 × 4.8	⎬ Interconnected	
	8	3.3	⎭	
Mordenite	12	6.7 × 7.0	0.26 ⎫ Interconnected	
	8	2.9 × 5.7	⎭	
Chabazite	8	3.6 × 3.7	0.48	
Analcime	6	2.6	0.18	
Zeolite ZSM-5	10	5.6 × 5.4	0.32	

[a] May not be the crystallographic measurements.
[b] Expressed as cm³ liquid H_2O/cm³ crystal.

35 and 36; the reader is also recommended to return to Fig. 2 which is a simple representation of chabazite.

Summary

It is hoped that the structural descriptions in this chapter convey the concept of zeolites as porous media often composed of a series of regular channels and cavities defined earlier. Access to these interstitial voids is via well-defined 'windows' composed of various numbers of tetrahedra (i.e. 'O' atoms) so that a well-constructed space of three-dimensional sieve is created. The dimensions of these channels and cavities are critical to the unique properties shown by zeolites. These will be considered in more detail in later chapters but Table VIII lists some critical dimensional parameters in common zeolites.

CHAPTER 4

Zeolite structure identification and characterization

X-Ray methods

Ideally all zeolite structures should be capable of solution by the use of modern X-ray single-crystal techniques and indeed naturally occurring species have been so studied. In the case of synthetic materials this requires the production of suitable single crystals. Despite special crystal growth techniques this is often not possible and has meant that a wide array of other techniques has been used for the investigation of zeolite structures.

Prominent amongst these is the sophisticated interpretation of X-ray powder diffraction data (XRD). In this method X-ray irradiation of zeolite powders (say $1-50\,\mu m$ crystallite diameter) produces a scattering pattern from the regular arrays of atoms (or ions) within the structure. It reflects the framework and non-framework symmetry of the constituents of each zeolite to produce a diagnostic fingerprint of 2θ (or 'd') spacings according to the Bragg equation:

$$n\lambda = 2d \sin \theta \qquad (4.1)$$

where $n =$ an integer, λ is the wavelength of the incident X-rays, d is the value of the interlayer spacings of the component atoms and ions and θ is the scattering angle.

These diagnostic patterns (Fig. 37) can provide an identification of known zeolite structures and Von Ballmoss has collected together a handbook (see Bibliography) of computer-generated 'standard' patterns. Examples of these are given in Table IX which shows the characterization of analcime and zeolite A via the 'fingerprint' of 2θ values and their relative intensities (i.e. the peak heights shown in Fig. 37).

This has an obvious use as quality control for known synthetic zeolites but must be used with extreme caution in claiming the synthesis of a new zeolite phase as syntheses often produce mixtures of phases (zeolitic and non-zeolitic) which require expert interpretation.

Recently major advances in computer-linked XRD data interpretation have given a new aspect to the use of power patterns for zeolite structural identification and prominent amongst these is the Reitvald method. In essence this

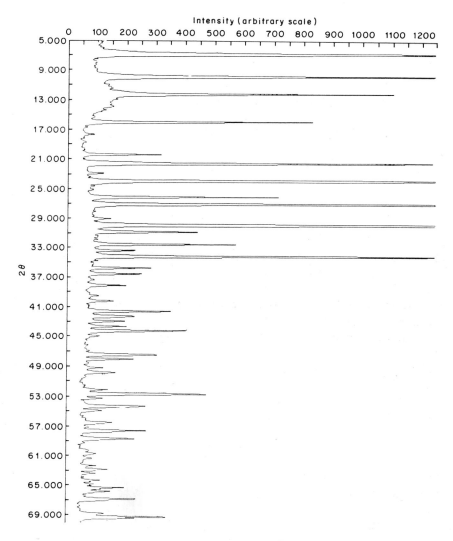

Fig. 37. An X-ray powder diffraction trace of zeolite A

predicts likely XRD patterns from simulated structures and presents data output of the closeness of fit between experimental and computed patterns. An example of this is given in Fig. 38. The method can predict framework atom positions as well as cation sitings and water molecule locations (to some extent).

Another widely used X-ray-based technique is that of X-ray fluorescence (XRF) analysis. This uses the X-ray-induced fluorescence emitted at wavelengths characteristic of atoms present in a solid as a quantitative analysis of elemental composition. The method requires very careful calibration and needs gram quantities for analysis. Although it is normally limited to atoms of atomic

Table IX. Simulated XRD powder patterns for analcime (Von Ballmoos, 1984)

(a) Analcime (ANA)

2θ	$d(\text{Å})$	I^a	2θ	$d(\text{Å})$	I
15.81	5.605	94.9	46.79	1.942	2.5
18.28	4.854	14.5	47.77	1.904	15.3
24.26	3.669	23.8	48.74	1.868	19.4
25.96	3.433	100.0	52.48	1.744	13.3
30.54	2.927	32.0	53.38	1.716	7.8
31.93	2.803	5.7	54.28	1.690	1.4
35.82	2.507	30.2	55.16	1.665	2.2
37.04	2.427	7.8	56.04	1.641	5.2
40.50	2.227	20.0	56.91	1.618	3.3
41.60	2.171	1.0	57.77	1.596	1.1
42.80	2.119	1.3	59.46	1.555	4.4
44.77	2.024	1.4	—	—	—

(b) Zeolite A (dehydrated) (LTA)

2θ	$d(\text{Å})$	I	2θ	$d(\text{Å})$	I
7.20	12.278	100	35.83	2.506	0.7
10.19	8.682	50	36.60	2.456	2.8
12.49	7.088	23.2	38.09	2.363	1.4
16.14	5.491	19.7	40.23	2.242	0.5
17.70	5.012	1.9	41.61	2.170	1.4
20.46	4.341	3.0	42.29	2.137	1.4
21.41	4.151	0.8	42.96	2.106	1.6
21.72	4.093	9.6	43.61	2.075	1.0
24.04	3.702	13.7	44.27	2.046	2.2
26.17	3.405	0.6	47.42	1.917	1.8
27.18	3.281	9.1	48.03	1.894	1.9
30.01	2.978	11.1	52.73	1.736	4.3
30.90	2.894	1.8	54.41	1.686	1.3
32.62	2.745	1.5	56.10	1.626	1.0
33.45	2.679	1.0	57.67	1.598	1.6
34.26	2.618	8.2	58.74	1.572	1.0

[a] I = relative intensity.

number greater than about 20 this does not greatly impair its use as a major approach to the quality control of synthetic zeolites.

Other diffraction techniques

Neutron diffraction studies afford several benefits in zeolite structural investigations. Samples can be in powder form and Reitvald analyses can be used. In addition the technique is intrinsically more sensitive to Si and Al ordering in the

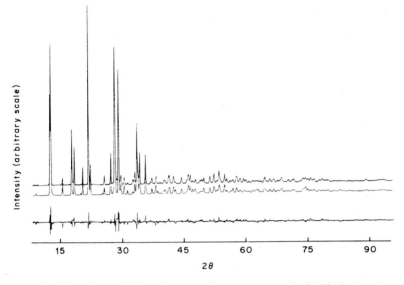

Fig. 38.   The Rietvald method applied to zeolite structure analysis. The bottom 'pattern' shows the difference between the recorded (top) and computer-generated (middle) diffractometer traces for gobbinsite. From L. B. McCusker and C. Baerlocher, *Zeitschrift für Kristallographie*, **171**, 281 (1985)

zeolite framework position as Si and Al have significantly different neutron scattering properties (but very similar X-ray scattering). Neutron diffraction gives a clearer identification of water molecule arrangements albeit that scattering from water molecules produces an inherently high background when other parameters are under investigation. An example of the details of water environments in zeolite structures revealed by neutron diffraction is shown in Fig. 39.

Electron diffraction is useful to zeolite crystallographers in that it clearly defines the faulted and twinned crystals common in synthetic and natural zeolites. It also detects the presence of superlattices and intergrowths. This is particularly important to the study of high-silica zeolites such as those in the ZSM series. The associated technique of scanning electron microscopy (SEM) is widely used in zeolite characterization. It is useful in quality control for the examination of new phases and mixed zeolite phases (Fig. 40). Obviously SEM displays crystalline habit and so interests the zeolite minerologists. Some examples of scanning electron micrographs of natural zeolites are given in Fig. 41.

High-resolution electron microscopy (HREM) gives even more exciting detail of zeolite crystal symmetries and has such a high resolution that one can 'see' zeolite pore structures (Fig. 42). It is the most appropriate way of examining defected and mixed phases of zeolite character (Fig. 43).

42

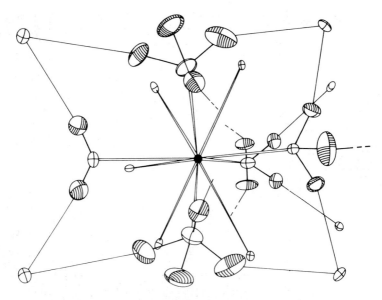

Fig. 39. A neutron diffraction analysis of water positions in brewsterite. The zeolite contains Sr (black circles) surrounded by H_2O (H—shaded circles. Open circles are both water and framework oxygens. Reproduced by permission of Elsevier Science Publishers BV from G. Artioli *et al.* in B. Držaj *et al.* (eds.), *Zeolites—Synthesis, Structure, Technology and Applications*, 1985, p. 252

Spectroscopic techniques

Inevitably infrared spectroscopy has found a very wide application to zeolite science. There are hundreds of papers to be found cataloguing research devised to illustrate the following:

(1) The nature of OH groups produced in zeolite structures, e.g. by calcination or by the presence of specific cations such as those of lanthanum and cerium.
(2) The quantification of catalytic sites (Brönsted and Lewis acid sites) by specific adsorbates (e.g. pyridine).
(3) Studies of species sorbed into zeolites—relating to framework sorbate and cation sorbate interactions for instance.
(4) Examinations of framework vibrations relating to the nature of sbus, effect of heat, presence of cations and sorbates etc.
(5) Study of zeolite surface hydolysis and metal aggregate deposition.

Recently Fourier transform infrared (FTIR) spectroscopy has been finding an increasing use in zeolite studies. The more exact relevance of these IR studies will become more apparent when zeolite catalysis is discussed in a later chapter.

UV and visible spectroscopy have been less useful to the zeolite chemist but some special studies of transition element complex ions have been carried out and luminescence spectroscopy is currently attracting interest as a means of following photochemical processes taking place in zeolite substrates.

Fig. 40.   SEM (× 88,000) showing zeolite A with zeolite impurities ('balls of wool').
Photograph: Laporte Inorganics, Widnes, UK

Resonance spectroscopy

Nuclear magnetic resonance (NMR) studies on zeolites have produced nearly as many papers as IR investigations. Its contributions can be summarized as follows:

44

Fig. 41. Natural zeolite morphologies illustrated by scanning electron microscopy. (a) Gmelinite with analcime ($\times 10.5$), (b) harmotome ($\times 21$), (c) chabazite ($\times 84$) and (d) clinoptilolite ($\times 84$). Reproduced by permission of Springer-Verlag from G. Gottardi and E. Galli, *Natural Zeolites*, 1985

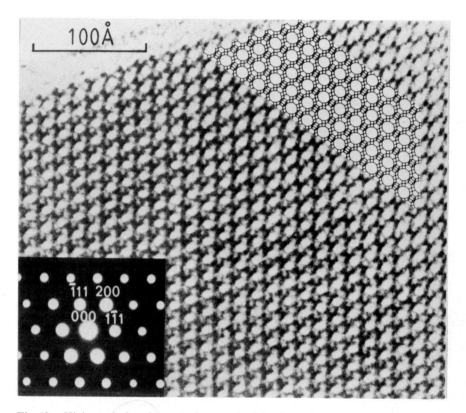

Fig. 42. High-resolution electron micrograph of the channel arrangements in NaY. Inset are the electron diffraction patterns and a schematic illustration of the Y framework. (Photograph Dr G. R. Millward, University of Cambridge)

(1) The use of spin echo techniques to quantify diffusional parameters of sorbates (via ^1H and ^{13}C resonances) and cations (e.g. Na, Li and Tl),
(2) The study of proton resonances to quantify H, OH and related species in zeolites and in zeolite crystal surface barriers,
(3) The use of magic angle spinning nuclear magnetic resonance (MASNMR) to elucidate the environments of the nuclei ^{29}Si, ^{27}Al, ^{13}C and ^{31}P in zeolite structures.
(4) The use of ^{129}Xe as a probe for characterization of zeolite sorptive properties.

Each of these areas of application has its own considerable contribution to make to the study of zeolitic properties but only one will be discussed in detail here. This is the use of MASNMR, which has a special place in zeolite structural and characterization exercises. The other NMR studies will be mentioned in later chapters with examples of their value.

46

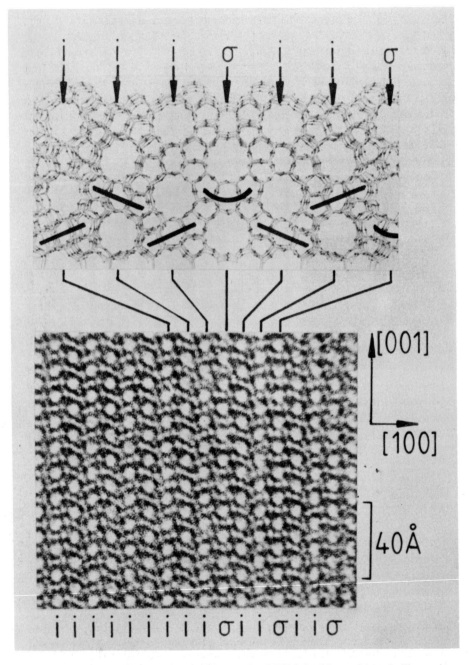

Fig. 43.   High-resolution electron micrograph of ZSM-5 with a schematic illustration (top). Three defect planes (σ) are seen interspersed with the normal planes of i symmetry (bottom). (Photograph Dr G. R. Millward, University of Cambridge)

MASNMR studies in zeolites

The rotation of the asymmetric nuclei required to produce nuclear magnetic resonances at specific ('magic') angles gives information as to the nearest coordination neighbours of the assumption nucleus. The only other technique producing similar information is that of Mössbauer spectroscopy, but that method is restricted to a very few nuclei (e.g. Fe, Sn) and its application to zeolites will not be covered here. The MASNMR observations give information on (i) differences between Si coordinated to 1, 2 or 3 other Si atoms or to various Al atoms (Fig. 44)—moreover it *quantitatively* defines the numbers of Si atoms in each different environment—and (ii) differences between Al tetrahedrally and octahedrally coordinated to oxygen atoms. The use of MASNMR to define Si environments requires the measurement of ^{29}Si MASNMR chemical shifts.

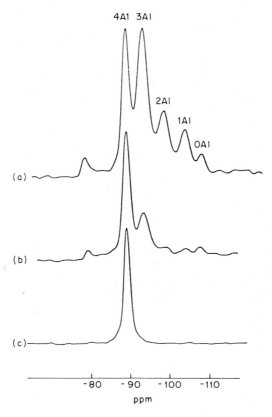

Fig. 44. Example of ^{29}Si MASNMR spectra for the different Si environments found in zeolite structures. Spectra are (a) ZK4 (isomorph of A) with Si:Al = 1.40; (b) ZK4 with Si:Al = 1. 13; (c) A with Si:Al = 1.0. Assignments correspond to Si with 4 Al as neighbours (4 Al), 3 Al as neighbours (3 Al), 2 Al as neighbours (2 Al) and 1 or 0 Al as neighbours (1 Al, 0 Al). Reprinted by permission from M. T. Melchior *et al.*, *Nature*, **298**, 455. Copyright © 1982 Macmillan Magazines Ltd

Table X shows how each geometry creates ranges of chemical shifts but the spectra produced are diagnostic (Fig. 44).

MASNMR has been important (with neutron diffraction) in defining the Si–Al ordering in aluminosilicate frameworks. Traditionally this is governed by the Löwenstein rule which excludes Al–O–Al linkages and MASNMR shows this to be true in zeolite frameworks with Si:Al greater than or equal to 1. The extent to which ratios > 1 show preferred local Si–O–Al, Si–O–Si environments (i.e. as in

Table X. Ranges of ^{29}Si MASNMR chemical shifts in zeolites[a]

Coordination	Notation	Chemical shift[b] (ppm)
Al | O | Al–O–Si–O–Al | O | Al	Si(4 Al)	$-86 \rightarrow -90.5$
Al | O | Al–O–Si–O–Al | O | Al	Si(3 Al)	$-88 \rightarrow -97$
Al | O | Al–O–Si–O–Si | O | Si	Si(2 Al)	$-93 \rightarrow -102$
Al | O | Si–O–Si–O–Si | O | Si	Si(1 Al)	$-97.5 \rightarrow -107$

(*contd.*)

Table X. (*Contd.*)

Coordination	Notation	Chemical shift[b] (ppm)
Si | O | Si–O–Si–O–Si | O | Si	Si(0 Al)	$-101.5 \rightarrow -116.5$

[a] Using tetramethylsilane as standard.
[b] Note these are typical values—some small variation between research groups is evident in the literature.

the first column in Table X) is not yet well defined although it seems to be non-random in some structures.

^{27}Al MASNMR spectra show clear differences with Al coordination (Fig. 45). This has provided the only satisfactory method for distinguishing between Al in framework sites (tetrahedral) and extra framework sites (octahedral). This definition has become important with the use of specialized treatments, both to remove framework Al from zeolites, to create higher Si:Al ratios and to introduce Al into very high Si zeolites, for the purpose of modifying their catalytic properties. MASNMR, also, can monitor similar modifications involving the incorporation of phosphorous and gallium into zeolite catalysts.

Other techniques

Electron spin resonance (ESR) spectroscopy has also been used to study zeolites—specifically to follow their modification by transition elements and to examine products formed by irradiation of zeolites with and without sorbates present. These investigations are primarily of a research nature and do not find much use for characterization or structural studies, although ESR has proved very useful in identifying specific lattice sites occupied by such ions as Cu^{II} and Mn^{II}. Every conceivable surface analysis technique has been used to study zeolites. Detailed discussion is beyond the scope of this book but brief mention of two areas of study can be made.

The first is the use of surface analysis to differentiate between surface and bulk properties. This can be illustrated by the use of electron probe microanalysis which has demonstrated the existence of silicon-rich surface layers in ZSM-5. The other is that of the use of several techniques to determine the nature of metal aggregates formed in the bifunctional zeolite catalysts to be discussed later.

Another widely used zeolite characterization method is thermal analysis and this again will be discussed in detail in the appropriate chapter.

50

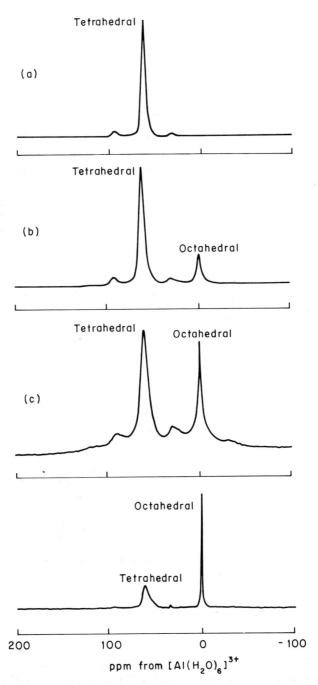

Fig. 45. ^{27}Al MASNMR for octahedrally and tetrahedrally coordinated Al. Diagram shows the increase of Al (octahedral) for a sample of NaY (a) with calcination (b), steaming (c) and prolonged steaming (d). Reprinted by permission from J. Klinowski *et al.*, *Nature*, **296**, 533. Copyright © 1982 Macmillan Magazines Ltd

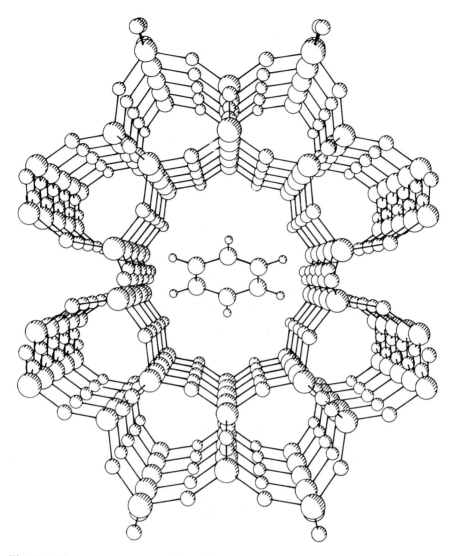

Fig. 46. The computer-generated graphic of a benzene molecule sorbed in the channel of zeolite Theta-1. Supplied by Dr S. Ramdas, BP

To close this chapter reference can be made to the spectacular advances being made in the use of theoretical calculations to produce computer graphics of zeolite structures and, in some cases, their properties. An example of this has already been seen in Fig. 43 where a computer-generated structure is compared to a high-resolution electron micrograph and Fig. 46 shows a further structure produced from theoretical calculations.

CHAPTER 5

Zeolite syntheses

Introduction

The first claim to have synthesized a zeolite dates from 1862 when Deville reported the production of levyne (levynite) by heating a potassium silicate with sodium aluminate in a sealed glass tube. Breck (see Bibliography) lists many similar early claims but those produced prior to the availability of XRD relied upon mineralogical identification (via optical and chemical analyses), often on microcrystalline products.

The first syntheses reliably characterized by chemical analysis, optical properties and XRD are those carried out by Professor R. M. Barrer and coworkers around 1940. The emphasis of this work was to synthesize species identical to naturally occurring zeolites—an objective achieved in many instances. Inevitably new phases were produced, either in admixture with known minerals or as distinct new products. Later work has clarified the identity of most of the phases discovered at that time including the production of a ZK–5 species not found in nature. However it is the work carried out in the Union Carbide laboratories (1949 on) culminating in the report by Milton, Breck and others of the synthesis of zeolite A (reported in 1956) that is generally regarded as the first example of the synthesis of a *completely characterized* new zeolitic structure unknown in nature.

Very significant in this work was the fact that it was carried out under mild hydrothermal conditions ($< 100\,°C$), at atmospheric pressure and with high water concentrations. Thus conditions were much closer to those of the formation of zeolites under natural sedimentary conditions rather than to their formation under the more extreme geological conditions of high temperatures and pressures noted in Chapter 1.

The last 30 years have seen many systematic studies of zeolite syntheses, both to generate new structures and to clarify their modes of formation in the laboratory and in nature. As computer predictions say that there are six million conceivable zeolitic structures it seems that studies will continue for a little while longer!

Generally the starting point for zeolite synthesis is crystallization from an inhomogeneous gel, created from a silica source and an alumina source, combined with water under high pH conditions generated by OH^- ion concentrations. Control of the $SiO_2:Al_2O_3$ ratio in this gel qualifies the final framework composition of the product and usually all the aluminium available is taken into the zeolite final composition (a notable exception to this is zeolite A).

How the rest of the available parameters can be manipulated to create different zeolites is a complex problem not yet fully understood. (Note that not all natural zeolites have been successfully synthesized in the laboratory.) More details of the factors involved will now be considered under the following headings: (i) components for synthesis, (ii) reaction variables for syntheses and (iii) kinetics and mechanism in zeolite syntheses.

Components for zeolite synthesis

Sources of aluminium

The majority of laboratory syntheses use metal aluminates to provide aluminium to the reaction mixture—most commonly sodium aluminate. Alternatively freshly prepared $Al(OH)_3$, or perhaps Al_2O_3 and $AlO(OH)$, are used. Some syntheses use aluminium alkoxides, aluminium salts (particularly aluminium sulphate) or even natural aluminium sources from glasses, sediments, felspars and felspathoids. Aluminium-rich industrial wastes have also been employed.

Zeolite production from the aluminium source is enhanced by the presence of the $[Al(OH)_4]^-$ moiety. Fig. 47 illustrates the probability of this species being present as a function of pH. The optimal pH range for successful synthesis is

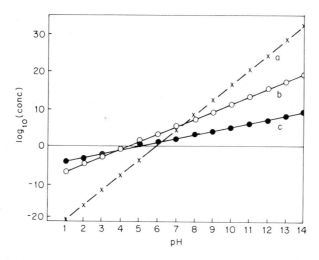

Fig. 47. Calculated concentrations of ratios of ion pairs for Al species in solution as a function of pH. (a) $Al(OH)_4^-/Al^{3+}$, (b) $Al(OH)_2^+/Al^{3+}$ and (c) $Al(OH)^{2+}/Al^{3+}$. (Data taken from R. M. Barrer, *Hydrothermal Chemistry of Zeolites*, Academic Press, London, 1982, p. 108.)

mainly pH 11–13, although some variation with zeolite species occurs outside this range.

Sources of silicon

The most widely used sources are soluble silicates and their hydrates (e.g. sodium metasilicate pentahydrate), silica sols (e.g. 30% by weight SiO_2) made from high surface area ('fumed') silicas such as Cab-o-Sil and commercial products (e.g. Syton, Ludox, etc.). Less frequently used are silica gels and glasses (including volcanic glass), silicon esters, clays (e.g. kaolinite), volcanic tuffs, sand and quartz.

Sources of cations

Obviously if high pH favours zeolite crystallization then alkali metals and alkaline earth hydroxides are the prefered cation source in most cases—certainly in industrial zeolite production. Other oxides and some salts have been used and, of course, the soluble silicates and aluminates quoted above are themselves cation sources.

Often mixtures of cations have been used and early work by Barrer and Denny (1961) used an organic cation—tetramethylammonium (TMA). Since their work many syntheses including organic cations have been developed and they are particularly useful in the syntheses of high-silica zeolites (e.g. ZSM–5), phases of silica with zeolitic properties (e.g. silicalite) and some zeotypes (e.g. the $AlPO_4$ materials).

Other materials in zeolite synthesis mixtures

Quite apart from organic bases many different organic compounds have profitably been included in syntheses mixtures. Some of these are listed in Table XIII, later in this chapter, but the list is by no means exhaustive, for instance a whole range of zeolites have been prepared from gels containing organic dyestuffs. The role that these materials and the cations play in crystallization will be discussed in the last section of this chapter.

Reaction variables in zeolite synthesis

Concentration, temperature, pressure and time

Usually reaction mixtures are composed of the appropriate sources to give the required Si:Al zeolite framework composition with an excess of hydroxide present. As the OH^-:SiO_2 ratio is increased more silicate remains in solution and lower Si:Al products are formed. Obviously the H_2O:SiO_2 ratio is inherently linked to this as well.

With regard to temperature, a synthesis of zeolite species at subambient temperature has been noted and it is certainly possible to prepare the gismondine-like synthetic species Na-P (NaP–I) at room temperature. As noted before most commercial preparations of A and X take place at 90–100 °C, at atmospheric pressures and under the conditions of high water activity conducive to formation of zeolites of high porosity (close to 50% of the framework of both A and X is available as void volume on removal of the synthesis waters—i.e. the 'zeolitic water').

Zeolites of lower water content have traditionally been prepared at higher temperatures (up to 350 °C) and pressures (developed autogeneously in sealed reaction vessels). Actually these more dense zeolite phases (e.g. analcime) can be favoured at high pH and increasingly this knowledge is being used to prepare even the denser zeolites under relatively mild hydrothermal conditions. Recent literature quotes low-temperature hydrothermal preparations of mordenite, ZSM-5 and analcime. These tend to be recipes using organic bases.

The element of time is important in two ways: (i) an induction period during which the reaction mixture is held near ambient temperature prior to raising to the crystallization temperature often optimizes zeolite yield (as in X, Y syntheses); (ii) often different zeolites crystallize from one reaction mixture at different times. This second time element is because all zeolites are metastable species and in nature many examples are known of diagenetic sequences whereby more open zeolite structures (e.g. phillipsite) convert, over a geological time span, to less open structures (e.g. clinoptilolite) and finally to analcime—the most stable and dense of common zeolites. On a laboratory, or plant, scale crystallization times are important to the production of A, X and Y which are metastable to NaP, so industrially reaction mixtures are quenched at optimal crystallization times. Similarly mordenite can transform to analcime.

These are examples of Ostwalds' law of successive transformation in which an initial metastable phase is transformed successively into one or more phases of higher stability. It has been found that an increased OH^- concentration can shorten crystallization times and Fig. 48 demonstrates this for a mordenite synthesis via the induction time. Part of this effect no doubt arises from an increase in the amount of reactants taken into solution with pH increase and hence the ensuing promotion of crystal formation.

This is not a simple effect, however, as Fig. 48 also shows that an increase to pH 13.3 drastically reduces yield by promoting the formation of the denser analcime phase as commented on earlier. Another aspect to pH control is that it can be a way of altering crystal morphology and an example of this is shown for silicalite in Fig. 49.

The correct combination of the variables discussed is critical if the formation of unwanted zeolitic and non-zeolite phases is to be avoided. Detailed crystallization fields have been worked out for several common reaction mixtures. Fig. 50 shows the products identified from gels of identical '$Na_2O.Al_2O_3$,' composition reacted at various temperatures with a variety of SiO_2 additions. The non-zeolitic phases created are albite (a feldspar), nosean, (a felspathoid) and paragonite (a micaeous layer silicate).

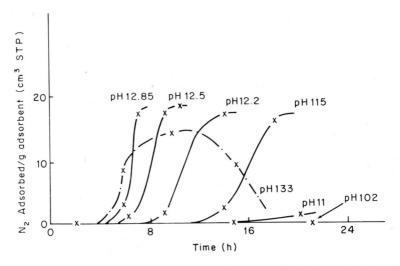

Fig. 48. Influence of pH on crystallization time for the synthesis of mordenite at 300°C. The N$_2$ sorption per gram of zeolite is used as a measure of the amount of mordenite produced. Reproduced with permission from R. M. Barrer, *Zeolites*, **1**, 133 (1981). Copyright © Butterworth Publishers

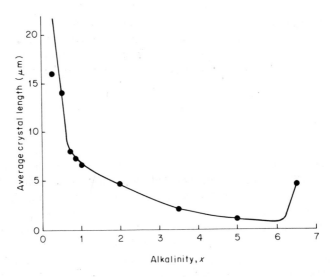

Fig. 49. Plot of the average length of silicalite crystals used to illustrate the changes in crystal morphology created by the variation of the alkalinity (x) of the reaction mixture. Reproduced by permission of the Royal Society of Chemistry from S. G. Fegan and B. M. Lowe, *J. Chem. Soc. , Faraday Trans.* 1, **82**, 785 (1986)

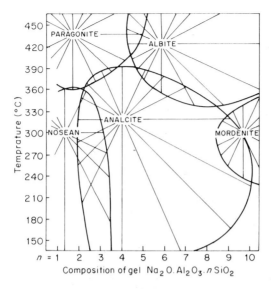

Fig. 50. Typical zeolite crystallization field showing the phases formed at different reaction temperatures from a sodium aluminosilicate gel of uniform composition. Nosean (a felspathoid) paragonite (a mica) and albite (a feldspar) are the non-zeolitic phases formed. Reproduced by permission of Academic Press from R. M. Barrer, *Hydrothermal Chemistry of Zeolites*, 1982

Another technique often used to aid zeolite production is that of 'seeding', i.e. the encouragement of the growth of a specific zeolite species by the addition of a small amount of this zeolite to the synthesis mixture. Alternatively the addition of another liquor from previous successful syntheses can be the source of 'seeds'.

Other parameters

Many other factors can influence zeolite synthesis. Several studies record the effects of small amounts of salt impurities (iron salts need to be excluded from synthesis mixtures) and the consequences of trace amounts of aluminium in silica sources. Others concentrate on the mode and nature of stirring and the influence of the order of addition of reactants together with the nature of the reaction vessels. Each factor *can* have an effect but no broad generalization can be drawn to help in the understanding of the mechanisms of crystallization involved, except to note that glass reaction vessels demonstrate memory effects and can drastically alter the course of syntheses. Presumably this comes from tiny seed crystals held in the glass surfaces etched by the high pH reaction mixtures. Plastic vessels or liners are advisable and even these must be very thoroughly cleaned before reuse.

Kinetics and mechanisms in zeolite crystallizations

Precursors

Currently there is no comprehensive explanation of routes whereby three-dimensional aluminosilicate structures grow from reaction mixtures to create zeolite frameworks. The most obvious missing step is an accurate knowledge of how different sbus arise from precursor species. Clearly the $[Al(OH)_4]^-$ species plays an important part, being a tetrahedral entity known to be present at high solution pH, but its reaction with silicate species to form aluminosilicate precursors to zeolite species is not clear. Perhaps the best evidence that such precursors exist is the observation that minute amounts of aluminium in the silicate sources used to make synthetic faujasites (i.e. X,Y) can be critical to their ease of production. Modern NMR and Raman spectroscopy have identified ring silicate and aluminosilicate anions in solution although information on zeolite-forming gels is sparse. This can be expected to be expanded as modern solid-state NMR and IR studies progress.

Given that precursors exist the next problem encountered is to explain the agencies whereby the precursors grow into differing structural frameworks, i.e. what are the structure-directing entities present in zeolite reaction mixtures? A

Table XI. Relationship between cation present and zeolite structure formed from reaction mixtures[a]

Zeolite	Cations present in synthesis[a]	Preferred cation
Gismondine	Na, (Na, TMA), (Na, Li), (Na, K), (Na, Ba), (Li, Cs, TMA)	Na
Gmelinite	Na, Sr, (Ca, TMQ), (Na,TMA)	Na
Faujasite	Na, (Na, TMA), (Na, Li), (Na, Ba)	Na
Zeolite A	Na, (Na, TMA), (Na, K), (Na, Ba), (Na, Ba, TMA), (Li, Cs, TMS)	Na
Mordenite	Na, Ca, Sr	
Edingtonite	K, Rb, Cs, (K, Na), (Na, Li), (K, Li), (Li, Cs, TMA), (Ba, Li), (Li, Cs)	K, Rb, Cs, Ba
Chabazite	K, Sr, (K, Na), (K, Li), (K, Ba), (K, Na, TMA)	K
Zeolite L	(K, Na), K, Ba, (Ba, K), (Na, Ba)	K, Ba
Yugawaralite	Sr, (Ba, Li)	Sr, Ba
Thomsonite	Ca	Ca
Epistilbite	Ca	Ca
Heulandite	Sr	Sr
Ferrierite	Sr	Sr
Zeolite ZK-5	Ba, (Na, Ba), (K, Ba), (Li, Cs, TMA)	Ba
Analcime	Na, K, Rb, Cs, H, NH$_4$, Ca, Sr, (Na, K), (Na, Rb), (Na, Cs), (Na, Tl), (K, Rb), (Rb, Tl), (Li, Cs)	None

[a] Bracketed ions refer to mixtures used in one synthesis recipe. TMA = tetramethylammonium

longstanding concept is that this arises from a templating action controlled by the cations present, as Table XI demonstrates. This corresponds, at least simplistically, to the accepted concept that there are two types of ion in solution—one which disrupts the existing water structure (structure-breaking) and one whose presence increases the existing water structure (structure-forming). Na^+ is an example of a structure-breaking ion and Ca^{2+} is a structure-forming ion. Another casual observation is that some sbus seem to be stabilized in synthesis by specific cations. Sodium ions have been said to promote structures containing D4R and D6R units, sodalite and gmelinite cages, whereas K, Ba and Rb promote cancrinite cages. However no similar comment can be directed to zeolite structures containing single ring sbus.

Templating arguments generated the original inclusions of organic cations into synthesis on the basis that a large cation might create larger channels and cavities to extend zeolite applications as catalysts and molecular sieves. The tetramethylammonium (TMA) cation is known to be strongly structure-directing and it is certainly retained inside the zeolite structure after crystallization and can often be observed in a special cage conformation. It is necessary to calcine zeolites so prepared before use as catalysts or sieves to remove the organic cation.

That incorporation of organic molecules into syntheses mixtures has been successful is not in doubt and Table XII shows some examples, including molecules which are not quaternary bases. However, it is still debatable whether or not this is a success of *templating*. Certainly model building can demonstrate closeness of fit between some cavity sizes and 'templates'. Other evidence comes from sodalite syntheses where TMA ions are sited one per cage and the Si:Al ratio is 5 as required for correct anion charge balance. As the TMA ion is too large to enter or leave the cage, the cages must form round the cation. A similar role of TMA in the sodalite cages for an A synthesis has been recorded. Again tetrapropylammonium ions (TPA) are sited at intersections of four channels in ZSM-5 with each of its four propyl groups directed along individual channels (Fig. 51).

On the other hand ZSM-5 can be synthesized *without* an organic template and can also arise from the incorporation of *many* organic molecules into appropriate synthesis recipes. This is true of other zeolites and also of the $AlPO_4$-5 zeotype. This is demonstrated in detail in Table XII. Furthermore, quite exotically shaped organic moieties still produce fairly mundane zeolites (Table XIII).

A natural conclusion to these points is that perhaps the organic presence promotes a modification to the gel chemistry. Some organics solubilize silica, via a simple complexation process (e.g. catechol) or via an increase in pH (e.g. amines and quaternary ammonium hydroxides), and thereby increase silica solubility as seen earlier. Alternatively some polymeric materials reduce dissolution to $Si(OH)_4$ by sorption onto silica surfaces. Recently direct reactions between organic bases and reaction mixtures to form organosilicate and organoaluminosilicate species have been recorded, including one whereby a D4R ion and a

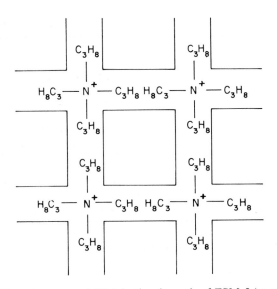

Fig. 51.   The orientation of TPA in the channels of ZSM-5 (as synthesized)

Table XII.   Multiple 'templating' action of various organic entities

ZSM-5 structures	AlPO$_4$-5 structures
Tetrapropylammonium	Tetrapropylammonium hydroxide
Tetraethylammonium	Tetraethylammonium hydroxide
Tripropylamine	Choline hydroxide
Ethyldiamine	Et$_3$N
Propanolamine	Bu$_3$N
Ethanolamine	(CH$_2$CH$_2$OH)$_3$N
Methyl quinuclide	Cyclohexylamine
NH$_3$ + alcohol	N,N'-Dimethylbenzylamine
Glycerol	Diethylethanediamine
n-Propylamine	Aminodiethylethanolamine
Di-n-butylamine	Dimethylethanolamine
Di-n-propylamine	Methyldiethanolamine
1,5-Diaminopentane	Methylethanolamine
1,6-Diaminohexane	2-Picoline, 3-picoline, 4-picoline
Morpholine	Diethylpiperazine
	DABCO[a]
Pentaerythritol	N-Methylpiperidine
Dipropylenetriamine	3-Methylpiperidine
Dihexamethylenetriamine	N-Methylcyclohexylamine
Triethylenetetramine	Dicyclohexylamine
Diethylenetetramine	Ethyl-n-butylamine
1-alkyl-4-azonibicyclo [2, 2, 2]	
octane-4-oxide halide	
Hexanediol	
Propylamine	

[a] DABCO = 1,4-diazobicyclo [2,2,2] octane (C$_6$H$_{12}$N$_2$).

Table XIII. Zeolite syntheses from reaction mixtures containing organic species.

Organic species	Zeolite formed
Tetraethylammonium	Mordenite, ZSM-8
Tetrapropylammonium	ZSM-5
n-Propylammonium	ZSM-5
Tetrabutylammonium	ZSM-11
Choline	Ferrierite-like structures (ZSM-38)
Pyrrolidine	Ferrierite-like structures (ZSM-35)
1,5-Diaminopentane	ZSM-5
1,8-Diaminooctane	ZSM-11
MQ[a]	Levynite-like structures (ZSM-20)
BP[b]	Losod
Neopentylamine	Mordenite

[a] MQ = methylquinuclidine ($C_8N_{16}N^+$).
[b] BP = bispyrrolidinium-5-azoniaspiro(4,4)nonane ($C_8N_{16}N^+$).

tetrabutylammonium ion react to produce five-membered siloxane rings. This is a critical observation because it is the first indication of a possible route to the formation of the five membered Si,O rings common to many zeolites (i.e. those containing 5–1 and 4–4–1 sbus). These tend to be in high-silica zeolites which are favoured in organic-based synthesis mixtures.

It is appropriate at this juncture to point out that water clathrates have pentagonal symmetry and that a complete elucidation of the structure of fully hydrated zeolite A shows that the water molecules inside the large cavities (α cages) are arranged in the pentagonal dodecahedral structure well known in clathrates (see Fig. 34). In this full elucidation no Na^+ was present at the centres of either the α or β cages. It therefore seems at least a possibility that a templating action might arise from a water structure created in the gel by cation influence (one or more cations), i.e. not a primary template from hydrated ions but an extension of cation influence to create a water structure, around which aluminosilicate frameworks could grow. Now the primary function of organic cations is that of accurate control of pH to maximize silica solubility, coupled with the promotion of structural precursors, rather than creating a former for structures. This tenet encompasses circumstances where *unchanged* organic molecules are added to zeolite-forming gels, thereby aiding the formation of specific zeolites, as it has long been known that organic molecules can attain a limited solubility in water via the formation of water 'clathrates'. It also explains the structure-directing effect that *anions* have—especially in felspathoid syntheses. Finally the fact that a polyelectrolyte (MW $\sim$ 10,000) containing the following units:

$$\left[- N^+ \bigcirc N^+ - (CH_2)_n - \right]$$

directs mordenite and gmelinite formation in a gel composition normally giving Y, NaP or analcime. Polyelectrolytes are used industrially to alter gel composi-

62

tions and an analogy might be drawn to the clathrate formed from urea by hydrogen binding round long n-paraffin chains in parallel one-dimensional channels. Perhaps the polyelectrolyte promotes the creation of the channel structures pervading the mordenite and gmelinite structures by a similar water clathrate as an 'overcoat' round the polymer chains. Cages form around the $-N-N-$ parts of the polymer and gmelinite forms best when $n = 4$ in the polymer sub-unit which then has a good match to the gmelinite cage size.

CHAPTER 6

Zeolites as ion exchangers

Introduction

The property of zeolites which allows the replacement of cations held in their aluminosilicate anion framework by ions present in external solutions or melts has been intensively studied for a variety of reasons. Most effort has been placed into zeolite 'modification' whereby the introduction of cations, usually by ion exchange, has been used to modify the catalytic or molecular sieving actions of the parent zeolite. Such is the potential of this modification that virtually every cationic form of the elements in the Periodic Table has been introduced into a zeolite framework.

Other investigations of zeolitic ion exchange have encompassed the limited (but useful) applications that they have in certain specialized areas and their importance as models for theoretical studies of the ion exchange process.

In view of this considerable interest it is worth pointing out that not all zeolites undergo facile cation replacement. Some zeolites with condensed frameworks (i.e. high density and low porosity) have limited and slow exchange properties. Examples of these are natrolite and analcime. It is appropriate to continue this chapter with an outline of ion exchange theory related to zeolite studies.

Ion exchange equilibria

Some basic theory

Exchange between cation A^{Z_A}, initially in solution, and B^{Z_B}, initially in a zeolite, can be written as follows:

$$Z_B A^{Z_A} + Z_A \bar{B}^{Z_B} \rightleftharpoons Z_B \bar{A}^{Z_A} + Z_A B^{Z_B} \tag{6.1}$$

where $Z_{A,B}$ are the valencies of the ions and the characters with a bar relate to a cation inside the zeolite crystal. Simple examples are:

$$Na^+ + \bar{K}^+ \rightleftharpoons \bar{Na}^+ + K^+ \tag{6.2}$$

for a uni–univalent exchange and

$$2\,Na^+ + \bar{Ca}^{2+} \rightleftharpoons 2\,\bar{Na}^+ + Ca^{2+} \tag{6.3}$$

for a uni–divalent exchange.

These stoichiometric reactions can be conveniently characterized by the construction of an ion exchange isotherm. This is a pictorial representation of the equilibrium concentrations of the respective ions in both solution and zeolite phases.

Clearly before making up such an isotherm it is necessary to ensure that ion exchange equilibrium has been reached, so some simple initial kinetic measurements must be performed. In open zeolites (e.g. A, X and Y with low framework densities) equilibrium will be reached in about a week for uni–univalent exchanges. With ions of higher valency it is not unusual for the period of time to be significantly longer and exchanges of even uni–univalent type can take months in the more dense zeolites to reach completion.

When the equilibration time is determined the isotherm can be constructed as follows. Weighed amounts of zeolites of known water content are placed in contact, in plastic containers, with solutions containing *both cations* (say A^{Z_A} and B^{Z_B}). These solutions contain varying proportions (known) of A^{Z_A} and B^{Z_B}. It is vital that the solutions, although having varied *relative* amounts of A^{Z_A} and B^{Z_B}, should be of a constant total *normality* (N). This condition of isonormality means that the total solution ionic strength remains constant in each individual zeolite/solution system. In a uni–univalent exchange isonormality is the same as isomolarity (viz. mol dm^{-3}) but when ions of different valency are involved in the exchange care must be taken to maintain the correct isonormal rather than isomolar conditions.

Solution and zeolite phases are shaken, at constant temperature and pressure, until equilibrium. It is advisable that the solution/solid volume ratio be not less than 20.

At equilibrium solution *and* solid phases should be analysed to determine the distributions of A^{Z_A} and B^{Z_B} between the phases. An isotherm can now be plotted which records the equivalent fraction of the entering ion in solution (A_S) against that in the zeolite $(\bar{A}_Z)$.

The equivalent fraction (A_S) of A^{Z_A} in solution is given by:

$$A_S = Z_A m_A / Z_A m_A + Z_B m_B \qquad (6.4)$$

where $m_{A,B}$ are the concentrations (mol dm^{-3}) of the respective ions in solution.

Similarly the equivalent fraction in the zeolite is:

$$\bar{A}_Z = Z_A M_A / Z_A M_A + Z_B M_B \qquad (6.5)$$

where $M_{A,B}$ represent concentrations of the ions in the solid phase.

Idealized isotherm shapes are shown in Fig. 52 and give pictorial indications of the relative preferences of the ion for the solution and solid phases. If the solid phase has equal preference for A and B then the isotherm would be a straight line joining A_S, $\bar{A}_Z = 0$ to A_S, $\bar{A}_Z = 1$ (dotted line on Fig. 52). It can then be seen that isotherm (2) in Fig. 52 illustrates the case where A is selectively taken up by the zeolite, unlike isotherm (3) where A remains in solution whilst ion B is preferred by the zeolite.

The selectivity of the zeolite for an ion (A) can be expressed quantitatively as a

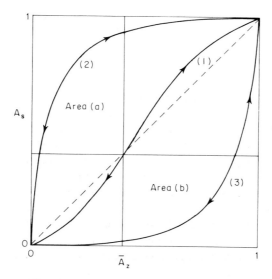

Fig. 52. Idealized ion exchange isotherms

separation factor (α) where

$$\alpha = \bar{A}_Z m_B / \bar{B}_Z m_A \tag{6.6}$$

(note: by definition $\bar{B}_Z = 1 - \bar{A}_Z$)

This can be measured graphically as area (a) divided by area (b), shown for a typical isotherm (1) in Fig. 52 and now when $\alpha > Z_A/Z_B$ zeolite is selective for A^{Z_A}, when $\alpha = Z_A/Z_B$ zeolite shows no preference and when $\alpha < Z_A/Z_B$ zeolite is selective for B^{Z_B}.

Significance of the interpretation of zeolite ion exchange isotherms

Although the use of ion exchange isotherms to quantify selectivity is a useful excercise it does not have a great significance *per se* because zeolites are only used as ion exchangers industrially in certain very restricted areas and very rarely as analytical ion exchangers.

Of wider inverest is a deeper consideration related to the understanding of the fundamentals of the ion exchange process. Zeolites have made a major contribution towards an understanding of the very complex interactions between the framework of an exchanger, the ions held within it, their associated solvent molecules and the properties of the external electrolyte solution in contact with the exchanger.

Initially theoretical studies devoted to organic-based ion exchanger resins and the clay minerals were intrinsically difficult to interpret because the basic ion exchange reaction was complicated by the simultaneous changes caused by the

66

swelling of the exchanger phase, sorption of non-electrolytes and by salt imbibition.*

Zeolites have the critical advantages as model cation exchangers in that their swelling is not significant and salt imbibition is usually very small under the concentration conditions commonly used to study their exchange properties (viz. $< 1N$).

The situation with regard to the influence of a non-electrolyte is more problematic and can have a very considerable effect on zeolitic ion exchange. This is illustrated by the observation that the presence of small concentrations of ammonia in solution can cause dramatic increases in cation exchange capacity, presumably because the zeolite prefers the complex ions created by the presence of the ammine ligand ($-NH_3$) over the originally uncomplexed ions in solution. However, these systems can be better studied in zeolites (rather than in resins and clays), simply because of the detail to which the structures of the three-dimensional zeolite framework, and its component cations and water molecules, have been quantified. This greatly improves the chances of an understanding of the basic processes involved.

This accurate knowledge of structure is a key to their importance as model cation exchangers, but it is an appropriate time to emphasize that, in order to extend the theoretical interpretation of a zeolite (or any other) exchange, it must be shown beyond doubt that the exchange being investigated is completely reversible throughout the full exchange capacity. This is because the theoretical analysis is based upon the principles of reversible thermodynamics. An example of an experimental system demonstrating reversibility is shown in Fig. 53. A

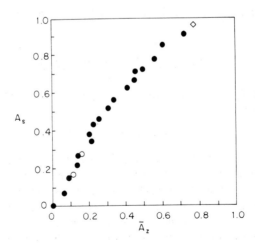

Fig. 53.   A reversible ion exchange, the $Na^+ \rightleftharpoons 1/2 Sr^{2+}$ exchange in phillipsite; Forward exchange ●, reverse exchange ○, repeated points ◇ (total normality = 0.2)

---

* Salt imbibition is the uptake of neutral salt molecules into an exchanger which readily occurs especially when there is a high salt concentration in the solution phase.

detailed treatise on zeolite ion exchange theory is beyond the scope of this book but some simple illustrations of the interpretations possible are of general interest and will be discussed in the following sections.

Analysis of zeolite isotherms to yield thermodynamic functions

When the ion exchange process can be described with certainty as a reversible equilibrium a mass action quotient (K_m) can be defined, as with all chemical equilibria:

$$K_m = A_Z^{Z_B} m_B^{Z_A} / B_Z^{Z_A} m_A^{Z_B} \qquad (6.7)$$

From the thermodynamic equilibrium constant K_a can be obtained where

$$K_a = K_m \Gamma (f_A^{Z_B} / f_B^{Z_A}) \qquad (6.8)$$

where

$$\Gamma = \gamma_B^{Z_A} / \gamma_A^{Z_B} \qquad (6.9)$$

$\gamma_{A,B}$ represents the single ion activity coefficients of the ions A^{Z_A} and B^{Z_B} in solution, and $f_{A,B}$ are the activity coefficients of the same ions in the crystal phase.

In practice K_a can be determined from the graphical integration of a plot of $\ln K_m \Gamma$ against $\bar{A}_Z$ (or by analytically integrating the polynomial which gives the computed best fit to the experimental data).

The quantity $K_m \Gamma$ often is described by

$$K_C = K_m \Gamma \qquad (6.10)$$

where K_C is the Kielland coefficient related to K_a by the Gaines–Thomas relationship:

$$\ln K_a = (Z_B - Z_A) + \int_0^1 \ln K_C dA_Z \qquad (6.11)$$

Having obtained K_a the standard free energy per equivalent of exchange ($\Delta G^\ominus$) can be found from

$$\Delta G^\ominus = - (RT \ln K_a)/Z_A Z_B \qquad (6.12)$$

where R and T have their usual meaning. The values of $\Delta G^\ominus$ obtained are more accurate quantifications of zeolite cation selectivities than α values. The $\Delta G^\ominus$ values can be used to generate figures for $\Delta H^\ominus$ and $\Delta S^\ominus$, the changes in enthalpy and entropy respectively, for the ion exchange process being considered. Evaluation of $\Delta H^\ominus$ requires definition at different temperatures, which means that interpretations using $\Delta H^\ominus$ values should be made with great caution as the levels of ion exchange in a zeolite commonly change with temperature. An example of this is the exchange of Na^+ for La^{3+} in zeolite X, illustrated by Fig. 54, which shows that exchange for Na^+ by La^{3+} is limited to about 80% (i.e. $\bar{La}_Z \cong 0.8$) at 25 °C but can be significantly increased by raising the temperature to 82 °C. Obviously it would be incorrect to determine $\Delta H^\ominus$ for the

68

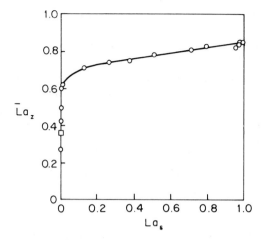

Fig. 54. The $Na^+ \rightleftharpoons 1/3\,La^{3+}$ exchange isotherm at 25°C for zeolite X. (The vertical axis gives the concentration of La in the zeolite and the horizontal axis gives the La concentration in solution). Reprinted with permission from H.S. Sherry in *Molecular Sieve Zeolites*—1, E. M. Flanigen and L. B. Sand (eds.), ACS Symposium Series 101; 1971, p. 357. Copyright (1971) American Chemical Society

$Na^+ \rightleftharpoons 1/3\,La^{3+}$ exchange in X at 25–82 °C as the standard state needed for the application of a correct thermodynamic analysis will be changing during the temperature range because the total number of ions involved in the exchange process has increased with temperature.

In conclusion it can be noted that $\Delta S^\ominus$ values are related to changes in water content created during the exchange. These changes arise when cations with different hydration spheres move between the solution and solid phases and/or when, for example, two univalent ions in a zeolite phase are replaced by one divalent ion creating a water flux to fill the empty space.

Selectivity and ion sieving

Selectivity series have been composed for several zeolites usually for the alkali and alkaline earth metals when the reversibility criteria are fulfilled. So far as the alkali metals are concerned, the series observed for the synthetic faujasites will suffice to illustrate the broad principles involved. The series for zeolite Y is

Cs > Rb > K > Na > Li,

whereas that for zeolite X is

Na > K > Rb > Cs > Li.

It was mentioned in Chapter 3 that the synthetic zeolites X and Y have identical anion frameworks differing only in framework charge. This difference is created by the relative amounts of aluminium isomorphously substituted for silicon in the aluminosilicate framework such that Y had a higher Si:Al ratio than X.*

Three conclusions arise from these two series viz: (i) zeolite X, with high anion framework charge (Si:Al $1 \rightarrow 1.5$) prefers the univalent alkali metals in the order of decreasing size; (ii) zeolite Y with a lower anion framework charge (Si:Al $\rightarrow 3$) prefers the alkali metals in the order of increasing size; (iii) the Li^+ ion is an exception to these two conclusions.

When similar analyses are attempted for the divalent alkaline earth metals no such simple pattern can be seen. Complications arise from two major sources, firstly the divalent ions have higher energies of hydration so that the size of the *hydrated* ion becomes important (as Li^+) and secondly it seems clear that the extent of exchange varies with cation at room temperature. Luckily the cation sites in X and Y have been determined so it is possible to correlate these to the extent of ion exchange in some cases.

The hydrated sodium form of zeolite X (NaX) can be taken as a convenient example. Each unit cell contains 16 Na^+ ions located in the hexagonal prisms of the framework (see Fig. 33) at S_I sites (one per prism) and a further 32 are in sites adjacent to the six oxygen windows (S6R) giving access to the large cages (in sites S_{II} and S'_{II}). The remainder of the cations are held in less well-defined sites within the large cage. When Ca^{2+} is the exchanging ion, complete replacement of the Na^+ ion occurs at room temperature (Fig. 55), but Ba^{2+} can replace only $\sim 80\%$ of the Na^+ at the same temperature. This percentage replacement corresponds to

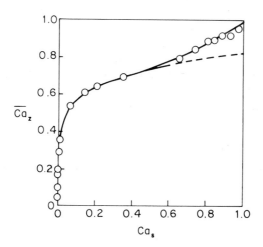

Fig. 55. The $Na^+ \rightleftharpoons 1/2\,Ca^{2+}$ exchange isotherm, at 25°C, for zeolite X. The dashed line shows the limit of the $Na^+ \rightleftharpoons 1/2\,Ba^{2+}$ exchange under similar conditions. Reprinted with permission from H. S. Sherry in *Molecular Sieve Zeolites*—1, E. M. Flanigen and L. B. Sand (eds.), ACS Symposium Series 101; 1971, p. 359. Copyright (1971) American Chemical Society

* In fact Y can be prepared within a range of higher Si:Al ratios than X up to about Si:Al = 3.

an inability to replace those Na^+ ions occupying S_I' positions in the unit cell. When the $Na^+ \rightleftharpoons 1/2\, Ba^{2+}$ exchange is carried out at 50° C all the Na^+ ions can be replaced from the zeolite. It can be concluded that the ambient replacement of Na^+ by Ba^{2+} experiences a limit, due to the large Ba^{2+} ion being reluctant to pass through the S6R at room temperature.

This provides an illustration of ion sieving inside a zeolite and is a common element to zeolite ion exchange. Another example is the exclusion of large cations from the sodalite cages in zeolite A and, of course, the limit to the $Na^+ \rightleftharpoons 1/3\, La^{3+}$ exchange mentioned earlier. This latter example, however, is the ion sieving of a highly hydrated cation (La_{hyd}^{3+}) from the hexagonal prism sites. A similar argument can be put forward to explain the anomalous selectivity for Li shown by X and Y. As Li^+ is known to be a highly hydrated ion in solution and rarely loses its solvent shell, it seems that its effective size approaches that of Cs^+ (unhydrated) and it is difficult to incorporate into the faujasite framework as the solvent shell shields the cation charge from the attractive negative framework charges. Another variant on this theme comes from the Mg^{2+} cation which is hydrated but does enter the X and Y structures through the restricting S12R. On the other hand zeolite A has a low preference for Mg_{hyd}^{2+} because the entrance of the ion into the A supercage is now controlled by the S8R which exerts an ion sieve effect. This steric exclusion of a 'complex' cation is also shown when zeolite A is found to accept simple monoalkylammonium ions but to largely exclude branched-chain monoalkylammonium and tetramethylammonium cations (Fig. 56). The same phenomenon can be demonstrated when ions in solution are either (i) solvated with molecules other than water or (ii) complexed by bulky ligands.

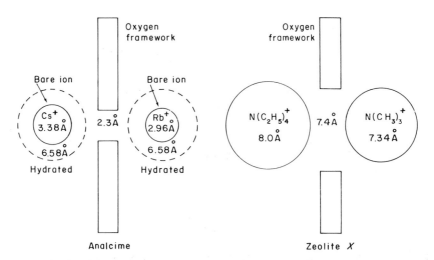

Fig. 56. Schematic representation of ion sieving. Cs and Rb are excluded by analcime (left) as both hydrated and non-hydrated species. The trimethylammonium ion(right) exchanges into zeolite X, but the tetrathylammonium ion does not

Although these examples have been compared to the synthetic zeolites X, Y and A they provide the basis for some general statements on the parameters controlling zeolite cation exchange properties, i.e.:

(1) The nature of both the competing ions with respect to their relative sizes and to their states of solvation inside and outside the zeolite.
(2) The charge on the zeolite framework coupled with framework geometry.
(3) The heteroenergetic nature of the cation sites available for occupation inside the framework.
(4) The temperature at which exchange is carried out, this can influence the removal of water of hydration, and the accessibility of sites and improve exchange kinetics.
(5) The concentration of the external solution coupled with the presence (or absence) of ligands other than water molecules.

Hysteresis in zeolite exchange isotherms

Several cases are known where, although the ion exchange in a zeolite is reversible, the reverse isotherm does not follow the same profile as the forward process. This is called 'hysteresis' and has been identified with the formation of two coexisting zeolite phases. The earliest known example is in the $Na^+ \rightleftharpoons K^+$ exchange in analcime (Fig. 57) where the hysteresis arises from the creation of a leucite phase at about 30% $Na^+ \rightleftharpoons K^+$ exchange (leucite is a potassium aluminosilicate phase of identical composition to analcime but with no water in the framework). Another example occurs in the $Na^+ = 1/2\,Sr^{2+}$ exchange in X

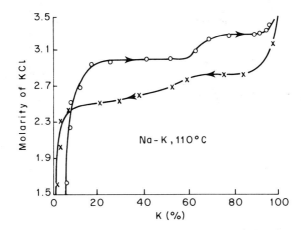

Fig. 57. The ion exchange isotherm for the $Na^+ \rightleftharpoons K^+$ exchange in analcime showing the hysteresis caused by the formation of a leucite. phase; Forward exchange ○, reverse exchange ×. Reproduced by permission of the Royal Society of Chemistry from R. M. Barrer and L. Hinds, *J. Chem. Soc.*, 1885 (1953)

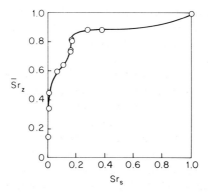

Fig. 58.   The $Na^+ \rightleftharpoons 1/2 Sr^{2+}$ exchange isotherm in zeolite X. Reprinted with permission from H. S. Sherry in *Molecular Sieve Zeolites*—1, E. M. Flanigen and L. B. Sand (eds.), ACS Symposium Series 101; 1971, p. 361. Copyright (1971) American Chemical Society

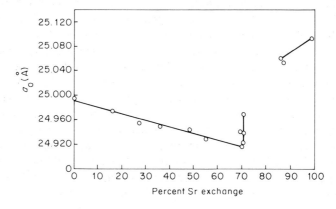

Fig. 59.   The change in unit cell dimension (a_o) with extent of $Na^+ \rightleftharpoons 1/2 \, Sr^{2+}$ exchange in zeolite X. Reprinted with permission from H. S. Sherry in *Molecular Sieve Zeolites*—1, E. M. Flanigen and L. B. Sand (eds.), ACS Symposium Series 101; 1971, p. 360. Copyright (1971) American Chemical Society

where a peculiar isotherm shape is identified with a new phase confirmed by XRD (Figs. 58 and 59). Hysteresis also arises when the coordination state inside a zeolite is significantly different from that in solution due to the presence of a non-electrolyte.

Hydrolysis and related phenomena in zeolites

In early ion exchange equilibria studies quite significant differences in values of $\Delta G^{\ominus}$ measured in apparently identical systems were noted. As even quite small errors in analysis can lead to quite dramatic changes in the shape of the Kielland plots (i.e. $\ln K_c$ versus $\bar{A}_z$) later work recommended (as herein) the analysis of *both* ions in *both* phases. This precaution should remove inconsistencies in $\Delta G^{\ominus}$ values, but in several well-documented cases this very careful analysis shows up further complications to complete ion exchange elucidation. In zeolites A and X, for example, it has become clear that hydronium ion exchange (i.e. $Na^+ \rightleftharpoons H_3O^+$) can play a major role in even 'simple' exchanges like that of Na^+ for Ca^{2+} (in A). Even in high-silica zeolites (mordenite, ZSM-5 and Y) this phenomenon has been detected and in exchanges involving transition element ions the effect of hydronium ion is so important as to require very careful buffering if true binary ion exchange is to be followed. Lack of accurate pH control leads to hydronium exchange, leaching of aluminium from tetrahedral framework sites, precipitation of metal hydroxy and oxide species onto zeolite crystal surfaces and ultimately to loss of zeolite structure.

Hydronium ion exchange is so facile in A that it can easily arise from excess washing of the zeolite, even with deionized water, which promotes the following zeolite surface reaction:

$$H_2O + Z-OH^- = H_3O^+ + Z-O^-$$

where Z represents the framework which has a high natural surface pH. (Remember that A and X especially are prepared at pHs as high as 13–14).

Recognition of the problems inherent to the interpretation of multicomponent exchanges have promoted major advances in ion exchange theory. This affords the means whereby the consequences of hydronium ion exchange superimposed on a binary exchange can be quantified, as can the $Ca^{2+}/Mg^{2+}/Na^+$ ternary system important to the use of zeolites as components in detergents (see below).

It is beyond the scope of this book to deal with the subtleties of these advances, but it is important to say that they now enable predictions of binary exchanges to be made and provide a theoretical framework to cope with exchanges involving any numbers of ions. They have also established fundamental links between ion exchange equilibria occurring in clay minerals and those in zeolites and are hence a basis for all inorganic exchangers (if not for resins).

Ion exchange kinetics

The problems outlined earlier, which are inherent to the theoretical interpretation of ion exchange equilibria, are equally applicable to an understanding of ion exchange kinetics. Here again zeolites have been useful models to try to improve existing theory. If anything, the need here is all the greater because most industrial uses of ion exchange depend critically on kinetics.

In a zeolite ion exchange the rate-controlling step is the so-called particle diffusion process. This means that the rate of the exchange depends upon the diffusion of cations, water molecules or cation–water complexes, through the zeolite framework.*

* At very low external cation concentrations ($< 10^{-5}$ M) the rate-controlling step to zeolite cation exchange *can* shift to the solution phase. Now the rate of exchange is dependent upon progress through the Nernst layer of orientated water molecules close to each zeolite particle surface. This is called 'film diffusion' and is easily recognized as now the rate of diffusion is affected by the speed of agitation].

Kinetic analyses of exchange rates can be built up via an appropriate solution to Fickian diffusion laws. There are several solutions available, of varying accuracy and complexity, but they all generate ionic diffusion coefficients from experimental data gathered under isothermal conditions. An example of this is seen in Fig. 60 which shows the rates of attainment of equilibria for the exchange between radioactively labelled cations inside zeolites and identical, but un-labelled, cations in solution at various temperatures. Each diffusion profile can be analysed to give a value for the self-exchange rate of Ca^{2+} expressed as the self-diffusion coefficient (D^*) for moving through the A framework. If a plot of $\log D$ versus $1/T(K)$ is constructed, application of the Arrhenius equation

$$D^* = D_0\, e^{-E_a/RT} \qquad (6.13)$$

enables calculation of a value for the energy of activation (E_a) of the self-diffusion process. D_0 is a pre-exponential factor representing the hypothetical self-diffusion coefficient at absolute zero.

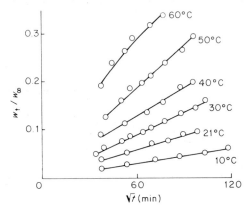

Fig. 60.   Fractional attainment of equilibrium (W_t/W_∞) as a function of root time ($t^{1/2}$) for the Ca self-exchange in zeolite A. (Followed by radioisotopic exchange.)

This E_a value represents the energy barrier to ion movement and can be used to suggest which route the ion may take through a zeolite framework when alternative routes, circumscribed by different oxygen windows, exist. It also can indicate when movement is likely to be that of an hydrated or unhydrated species. Fig. 61 shows a plot of energy barriers encountered by various ions moving through the zeolite A framework. Here the rate-determining step seems to be the progress through the S8R into the supercage; ions such as Ca^{2+} and Sr^{2+} shed waters to enter and so the height of the energy barrier encountered is a function of bare ion size. The Mg^{2+} ion retains its hydration shell and hence the barrier is unusually high.

Sometimes self-exchange rate plots can be related to the cation sites contributing to the process. An example of this is shown in Fig. 62 where the various

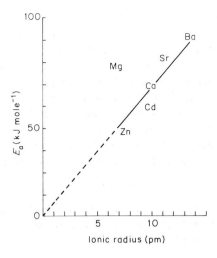

Fig. 61.   Variation of activation energy (E_a) with cation radius for ions moving through the zeolite A framework

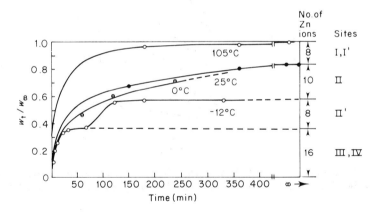

Fig. 62.   Fractional attainment of equilibrium (W_t/W_∞) for the self-exchange of Zn^{2+} in zeolite X at different temperatures. Figures on the righthand axis show numbers of Zn^{2+} (radiolabelled) ions replaced and related to their suggested sites within the X framework (see Fig. 33 for locations of the sites)

stages recorded for Zn self-exchange in zeolite X can be related to participating sites.

The application of similar analyses to heteroionic systems is less successful and although much progress has been made to the fitting of zeolite ion exchange kinetics to current theory, there is still some way to go. This does not prevent useful data being collected for heteroionic systems important to zeolite application as in the Na/Ca/Mg A system important to detergency.

Uses of zeolites as ion exchangers

Introduction

Although zeolite ion exchange has been well studied, often as an adjunct to their uses in catalysis and molecular sieving, very few practical uses have emerged.

A major barrier to zeolite use as ion exchangers is in their apparent lack of compatibility with column use. Synthetic zeolites normally crystallize in the particle size range $0.1 \rightarrow 10\ \mu$, which is too small to allow a reasonable liquid transport through a bed of particles. Although synthetic zeolites are usually marketed as crystal compacts (either self-bonded, by self-forming the required compact size from a performed gel, or bonded by using a clay binder) the compacts are not especially attrition resistant under liquid column use. They are also prone to damage via the hydrolytic processes mentioned earlier. This lack of compatability is much less evident with some naturally occurring zeolites and clinoptilolite is increasingly being used in quite large columns. The economics of the use of natural zeolites in this way at present depends critically upon their availability and whereas quite extensive use of clinoptilolite (and sometimes mordenite) columns is made worldwide, the source of the minerals is usually geographically close to its end use.

Another major reason preventing the wide use of zeolites as ion exchangers is their inherent instability in low pH. Zeolites with Si:Al ratios in the range 1–2 readily lose Al from their framework in acid environments with consequent loss of capacity and ultimate framework collapse. The limit of acid resistance of these zeolites is usually thought of as about pH 3–4. This is the general view of zeolite acid instability although it has been shown that zeolites with higher Si:Al ratios are much more resistant to acid attack and can retain useful capacities even after exposure to concentrated acids over long periods. An example of this is shown in Fig. 63.

Such uses that do exist arise where there are requirements for highly specific ion exchangers of usable capacity—robust and with good thermal and/or radiation stabilities. In these circumstances it is fair to say that resins offer little competition and that in general terms zeolites are cheaper than resins. The most important of modern uses of zeolites as ion exchangers will now be described.

Zeolites as components in commercial detergent composition

Some 20 years ago a growing awareness of the environmental damage created by the use of polyphosphates in detergents caused detergent manufacturers to seek less hazardous replacements. The function of the polyphosphate was to 'build', i.e. enhance, the cleaning efficiency of detergents primarily by removing Ca^{2+} and Mg^{2+} ions from washing water to prevent their precipitation by surfactant molecules. The use of zeolites as builders was suggested in the 1970s and since then it has been demonstrated that they can effectively carry out the

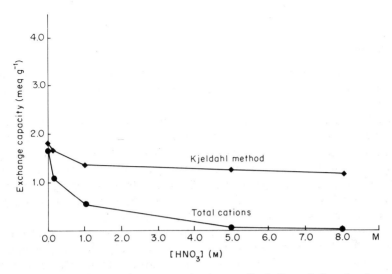

Fig. 63. Variation of the exchange capacity (meq g^{-1}) of clinoptilolite after long-term exposure to nitric acid solutions. Capacity measured by (a) NH_4 uptake (Kjeldahl method) and (b) by total ions lost by the acid treatment. Note that difference in capacity is due to H_3O^+ exchange

water softening exercise needed for successful laundering. All readily available zeolites (natural and synthetic) have been screened as 'builders', but the detergent industry has chosen zeolite A as being the most effective. At present the chemical turnover for A as a builder runs in hundreds of millions of pounds and to date at least two plants have been purpose built to cater for an annual demand estimated at over 500 million tons per annum in 1989.

Certainly Zeolite A as synthesized in its sodium form has the appropriate selectivity for Ca^{2+} hardness removal, coupled with the fast exchange kinetics needed to function at normal washing temperatures and cycles. Zeolite A has less success in removing Mg^{2+}, as it has a poor selectivity (over Na and Ca) and unfavourable kinetics for the reasons stated earlier. Usually manufacturers add other 'builders' to remove Mg^{2+} —often the undesirable sodium tripolyphosphate—but the sodium form of zeolite X can provide a satisfactory alternative. Some detergent formulations include mixtures of NaA and NaX as builders.

Two interesting side-issues have evolved from zeolite detergent use. Firstly, manufacturers have developed A crystals within tight particle size specifications having a special crystal morphology. In this morphology the more usual cubic habit of A with 'salt-like' acute edges and corners has been modified to one with bevelled edges so that it is more readily lost from fabrics during rinse cycles. This is illustrated in Fig. 64. The second consequence of the use of 'builders' is that it has demonstrated that zeolite A has no harmful environmental or human health effects.

Fig. 64.   Scanning electron micrograph of A (× 8250) showing (a) normal morphology and (b) morphology with chamfered edges. Photograph: Reproduced by Laporte Inorganics, Widnes, UK

So far as environmental safety is concerned, tests have shown that A has little or no effect when released to the environment at levels related to its widespread use as a detergent component. The tests are summarized in Table XIV. The human health effects of zeolite A have also been extensively investigated. Little adverse data has been collected as can be seen from Table XV.

This is an appropriate place to comment generally upon zeolite toxicity.

(b)

Fig. 64 (*cont.*)

Zeolite A can be regarded as a non-hazardous material and other data suggests that clinoptilolite has a similar bill of health. However, it is clear that a fibrous form of erionite can be accused of promoting lung disease in the same way that other resistant fibrous materials of similar crystallite geometries do (i.e. asbestos minerals). This has been well documented for a fibrous habit of erionite in the Cappodocian region of Turkey. At the time of writing the evidence is that this hazard is confined to this specific location and zeolite occurrence.

Table XIV. Results of environmental tests applied to zeolite A for detergent use.

Test	Result
Deposition in sewer systems	No significant accumulation
Fate of zeolite during sewage treatment	As similar solid particles no effect on treatment efficiency, removed in same way as other solids
Effect on heavy metals present in sewage	Little effect—may even be beneficial, i.e. to 'fix' toxic metals
Presence in sludges	No adverse effects—could be beneficial as above
Toxicity in aquatic ecosystems	None noted
Creation of oxygen-deficient environments in natural waters (eutrophication)	None noted, unlike phosphate-based builders which are the major cause of eutrophication
Uptake of organic residues onto zeolite surface	Like other solids present
Ultimate fate	Zeolite A is metastable and ultimately hydrolyses to harmless solids

Table XV. Tests on zeolite A for human safety purposes

Test	Result
Acute toxicity	
LD_{50}	Essentially non-toxic
Percutaneous toxicity	Essentially non-toxic
Irritation of the eye	None
Irritation of the skin	Some—transient
Sensitization of the skin	None
Inhalation	No effects
Subchronic toxicity	
Oral	No effects
Percutaneous	No systematic effects
Inhalation	No effects
Metabolism	Hydrolysed and excreted
Teratology	No effects
Chronic toxicity	
Oral	No effects, non-carcinogenic at 0.1% in diet
Inhalation	No effects at 50 mg m^{-3} for 12 months or at 20 mg m^{-3} for 22 months

Zeolites for the treatment of liquid nuclear effluents

When spent nuclear fuel elements are removed from a nuclear reactor they are stored under water in 'ponds'. The water is of pH 11.4 with respect to sodium hydroxide and sodium carbonate. Pond storage is for long periods of time to

Table XVI. Examples of zeolite use in the treatment of nuclear waste

Type of effluent	Location	Zeolite used	Isotopes removed	Plant details
High-level radioactive waste	Handford Nuclear Lab, USA	Linde AW-500 (Chabazite)	^{137}Cs	300 ft^3 bed 3 × 10^6 gal charge
Purification from above	Hanford Nuclear Lab, USA	Large part mordenite	^{137}Cs	Full-scale plant
Process condensate waste water	Hanford Nuclear Lab, USA	Large part mordenite	^{137}Cs	Not known
Low-level waste water from fuel storage pond	National Reactor Station, Idaho, USA	Clinoptilolite	^{137}Cs ^{90}Sr	4 × 5.3 ft^3 columns 12,000 gal water/ft^3 zeolite
Evaporator overheads and waste water	Savannah River, S. Carolina, USA	Linde AW-500	^{137}Cs	9.2 ft^3 up to 76,000 gal
Waste water from fuel storage	British Nuclear Fuels Sellafield, UK	Clinoptilolite	^{137}Cs ^{90}Sr	4000 m^3 water/day in SIXEP plant

82

allow thermal and radiation cooling to take place prior to reprocessing. During this storage period a build-up of fission products (especially the radioisotopes ^{137}Cs and ^{90}Sr/^{90}Y) occurs in the pond water. This contaminated pond water is the major source of medium-level radioactive waste and must be decontaminated. Zeolites have been used for this and related purposes by the nuclear industry for many years and Table XVI contains a resumé of these applications with some indications of the scale of usage.

The UK use at the Sellafield plant of British Nuclear Fuels plc successfully employs a natural clinoptilolite from the Mud Hills deposit in California to remove Cs and Sr isotopes from its liquid effluents and has thereby greatly reduced the amount of radioactivity released to the Irish Sea.

Another nuclear use for zeolites arose from the Three Mile Island reactor accident in America. Here the commercial zeolites Ionsiv™ IE-96 and A-51 produced by the Union Carbide Company (Linde Division) were very efficient in removing ^{137}Cs from high-level activity water produced in the reactor as a consequence of the accident. Recently incorporation of clinoptilolite into the diet of sheep grazing on grass contaminated by the Chernobyl accident has reduced levels of Cs isotope uptake into the sheep.

The use of zeolites to treat radioactive effluent is attractive for reasons other than that they are compatible with the preferred method of storage and disposal of medium-level solid radioactive waste. Spent zeolites can be readily incorporated into a cement matrix prior to encapsulation in stainless steel drums. In this, and other nuclear environments, they seem to be very resistant to exposure to high doses of β, γ and neutron radiation. Zeolites would also provide effective 'backfill' and barrier materials in underground nuclear waste repositories where their function would be to act as absorbers of radionuclides should they be released accidentally or by corrosion from the repository.

These applications clearly illustrate the advantages that zeolites can have as very selective radiation-resistant ion exchangers capable of removing picogram quantities of radioisotopes from high concentrations of other ions. Table XVII illustrates this selectivity in pond water treatment by listing the relative concentrations of ions competing for uptake.

Table XVII. Typical ion make-up of 'pond' water (pH 11.5).

Ion present	Concentration
Na^+	100 $mg\,l^{-1}$
Ca^{2+}	1.5 $mg\,l^{-1}$
Mg^{2+}	0.7 $mg\,l^{-1}$
Cs^+	17.0 $\mu g\,l^{-1}$
Sr^{2+}	0.52 $\mu g\,l^{-1}$

Waste water treatment

There is a constant need to remove protein breakdown products from water. In regions of the world where clinoptilolite, mordenite, phillipsite and chabazite are available as a cheap, relatively pure commodities, they have been much used for the removal of ammonia and ammonium ions from aqueous effluents.

Clinoptilolite is the most often used—again demonstrating a particular zeolite use relying on a critical specificity for one ion over several more abundant ones. The scale of its use can be large and several designs of waste water treatment plants using zeolite columns have been produced. Table XVIII displays some details of these. Other more general treatments using clinoptilolite are listed in Table XIX and apart from these there are many records of small-scale plants associated with fish farming. It seems likely that this use of natural zeolites will

Table XVIII. Use of zeolites in waste-water plants

Location of plant	Zeolite used	Size	Regeneration
Rosemount, USA	Clinoptilolite	0.6 MGD[a]	Yes
Occoquan, USA Virginia, USA	Clinoptilolite	22.5 MGD	Yes
Tahoe-Truckee, California, USA	Clinoptilolite	6 MGD	Yes
W. Bari, Italy	Clinoptilolite Phillipsite	10 m³/h	Yes
Vae, Budapest, Hungary	Clinoptilolite tuff	50 m³/day	Yes
Toba, Japan	Clinoptilolite tuff	80 m³/day	Not known
Japan	Clinoptilolite tuff	500 m³/day (soap and detergent waste)	Yes

[a]MGD = million gal/day.

Table XIX. Application of clinoptilolite to treat liquid effluents

Use	Species removed (or treatment)
Boiler waters	Fe
Condensates from soda ash products	NH_4^+ and oils
Heating and power station waters	NH_4^+
Drinking water treatment	30,000 m³/day pilot plant
Nutrient recovery from sewage	240 m³/day plant
Textile waste (polyamide spinning)	Caprolactam and lubricants
Waste waters	Oil
Organochlorine production clean up	Trichloroethylene
Non-ferrous metal production	Zn, Cu

84

continue to expand. In addition, competition with the natural species for use in this way will increase from the synthetic zeolites F and W, produced by Union Carbide, which have higher exchange capacities for NH_4^+ than natural species and have already been used successfully in regenerable columns.

Before leaving this topic note should be made of the use of various tuffaceous materials for nuclear and waste water 'clean-up'. These materials are a fine-grained ash-like rock produced as a primary product of relatively recent volcanic activity. They are widespread in areas of Italy, where some tuffs have been found to contain as much as 60% zeolite (usually chabazite and phillipsite) and to be useful low-grade ion exchangers.

Use of zeolites for soil benefication and related uses

The Japanese use around 6000 tons per annum of clinoptilolite and mordenite to control soil pH, moisture content and manure malodour. In some areas of Japan this is a centuries old tradition. The control of pH is related to the ability of the zeolite to function as a slow-release agent to improve nitrogen retention in the soil. This has been demonstrated with grass and cereal crops. Horticultural applications generally use 5–10% incorporation by weight of clinoptilolite into growing media. The value of this has been shown in the production of, for instance, tomatoes, bell peppers, house plants and strawberries and other crops as can be seen from Table XX. Laboratory experiments have shown that zeolite A has a similar beneficial effect. Obviously ion exchange has a role in these phenomena which needs more clarification. The overall benefit seems clear, namely, adding a zeolite to soil adds a high-capacity selective exchanger to a medium in which these attributes boost an already present property of the soil by virtue if its clay mineral content. So far as adoption of zeolites as benefication agents worldwide goes it seems unlikely that this will be economic except where natural zeolites are in the proximity—as in Japan.

However, there are two circumstances where this pessimism may not be justified—firstly, it may well be economic and sensible to make zeolite additions to arid, semi-desert soils and, secondly, the currently decreasing cost of zeolite A may well bring it into consideration as an additive to horticultural growing media.

Table XX. Effect of soil conditioning by zeolite additions

Crop	Best yield index (control = 100)
Wheat	115
Eggplant	120
Apple	128
Paddy	117
Carrot	163

There is still a need for very careful ion exchange kinetic and equilibria studies at low levels of nutrient in concentration (e.g. for Na^+, NH_4^+ and Ca^{2+}). Very little information is available describing the properties of either natural or synthetic zeolites at low loading, but as such it does indicate clearly that unusual and interesting changes in selectivity and capacity can occur. The information in Fig. 65 shows some data for ferrierite (which resembles clinoptilolite). It can be seen that although ferrierite has high capacities for both K^+ and NH_4^+, both these ions are much less selectively taken up in the presence of quite small concentrations of competing ions. This is particularly unexpected for K^+ as the zeolite naturally occurs in a potassium-rich form. This provides an excellent example of one of the factors listed earlier as being important to the ion exchange

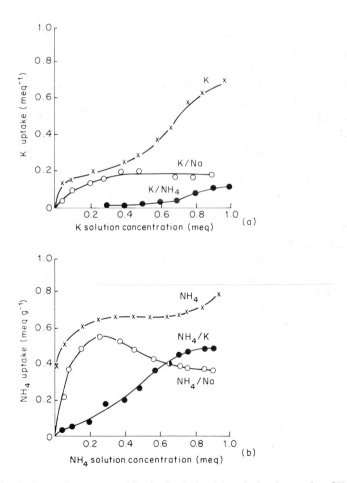

Fig. 65. Variation in ion exchange capacities for ferrierite; (a) variation in uptake of K^+ in the presence of Na^+ and NH_4^+ and (b) variation in uptake of NH_4 in the presence of K^+ and Na^+

process, i.e. that quite strange selectivities arise when specific geometric sites and certain ion pairs are controlling exchange in a special concentration range.

Zeolites as animal feed supplements

The beneficial effects of a 5–6% supplementation to the diet of pigs has been recognised in Japan for many years. Extension of similar studies to pig rearing in the USA, Cuba, Hungary, Austria, etc., has confirmed generally that pigs taking in clinoptilolite show beneficial weight gains and are less subject to disease than pigs fed by normal diets. The extent of the benefit varies between studies but all agree on the advantage. Part of the success seems to be derived from ion exchange control of ammonium ion levels in the gut, but some studies show higher levels of blood proteins, globulins and mineral elements in pigs receiving a Ca K form of clinoptilolite in their feed.

Supplementation of diet by zeolites in other creatures provides less convincing evidence of any benefits. Trials have been performed on goats, sheep, beef cattle, poultry, fish, rats and mice. These have nearly all used clinoptilolite, except for some recent careful studies in the USA which have used zeolite A additions.

It is difficult adequately to summarize these studies, but it is fair to say that in nearly all cases the creatures have benefited from dietary zeolite. Certainly there is no evidence of harmful effects, which is supporting evidence that zeolites are non-toxic. Criticism comes from some experts in animal husbandry and nutrition who say that the benefits seen reflect animal response to the special care allocated to them as a test group. Despite this comment it is known that addition of clinoptilolites to the diet of beef cattle is common practice in Hungary.

Here again the need is apparent for very detailed studies, especially of the physiological fates of the cations introduced via the zeolite into the animal host and of the ultimate chemical composition of zeolites passing through living systems. It is also possible that the presence of a porous high surface area material could affect host parasites (as has been demonstrated in rats), remove toxins or create changes in enzymology and immunological response. Obviously any of these effects could be synergistic. Data of this sort is being gathered slowly and one suspects that use of zeolites as dietary supplements for animals and fish will be confined to Eastern Europe, Cuba and Japan until more details emerge as to the nature of the processes promoted by zeolites in a living host.

CHAPTER 7

Zeolites as molecular sieves and drying agents

Introduction

Chapter 3 described the structural characteristics of zeolites, emphasizing their three-dimensional frameworks in which were enclosed a series of channels and voids. Contained in these interstitial zeolite volumes were the cations and water molecules.

When the water is removed the voids created within the framework can take in other molecules. This process is called 'sorption' and the zeolites are said to 'sorb' molecules into their void volume, i.e. they act as 'sorbents'. Frequently clathrate terminology is used, where the sorbing molecules are described as 'guests' within the zeolite 'host'.

Recourse to Chapter 3 also emphasizes the consequences of zeolite structural architecture, in that both the synthetic and natural minerals possess a wide variety of internal channel and cavity assemblages accessed through oxygen windows which happen to be close in size to the dimensions of common organic and inorganic molecules.

This geometry is the source of the ability of zeolites to separate mixtures of molecules (in both the gas and liquid phases) on the basis of their effective sizes—hence their description as 'molecular sieves'.

At this stage it is worth making two, very general, points. Firstly, it was mentioned on the first page of this book that other materials have molecular sieve properties (e.g. carbon and porous glass) but none have the potential flexibility of use that some zeolites do. Secondly, not all zeolites have conveniently accessible voids and not all zeolites lose water easily.

To further consider this second point note that Chapter 6 commented on the relative lack of cation mobility in zeolites with the more dense aluminosilicate frameworks. Similar strictures apply to molecular movement. If water migration is taken as an example, then the diffusion of water through analcime is much more hindered than through zeolite X. In zeolite X, the water migration is like that through liquid water. This is illustrated in Fig. 66 where the energy barrier (E_a) to water diffusion through various zeolites is plotted against a function of their framework density. Table XXI illustrates the same point showing that, for example, the diffusion of water through analcime is a factor of $\sim 10^6 \, \mathrm{m^2 s^{-1}}$

88

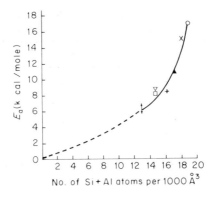

Fig. 66. Energy barriers (E_a) to water migration in zeolites as a function of framework density (Si + Al per 1000 Å³); analcime ○, natrolite ×, heulandite △, chabazite ▽, gmelinite □, NaP + and X,Y ↕

Table XXI. Diffusion coefficients (D^*) and energy barriers (E_a) for water in zeolite and water environments

Substrate	D^* (m² s⁻¹) (at 318 K)	E_a (kJ mol⁻¹)
Analcime	2.0×10^{-17}	71
Heulandite	2.1×10^{-12}	46
Chabazite	1.3×10^{-11}	36
Gmelinite	4.8×10^{-12}	34
Ice	1×10^{-14} (at 275 K)	56
Water	3.9×10^{-9}	19

slower than through chabazite and that the *flow* of water in a zeolite phase is often closer to that through ice rather than liquid water. This, and other data, has been used to suggest that zeolitic water has a unique ice-like state.

Returning to the special molecular architecture of zeolites it should be apparent that their single and double ring sbus offer selective separations of molecules on the basis of size. If these oxygen windows are assumed to be planar, the free diameters of various ring sizes can be calculated by assuming the component oxygens to have diameters of 2.7 Å. These are in summary in Table XXII.

Not all these dimensions are large enough to let molecules pass through them and so some zeolites do not act as three-dimensional porous structures so far as molecular penetration is concerned. Complications can arise when cations are sited close to the windows or in positions critical to molecular diffusion through the framework structures. These controls on molecular uptake into cavities and channels will be discussed later in this chapter.

Table XXII. Estimated diameters of
planar configurations of n oxygen atoms

n	Diameter (Å)
4	1.1
5	1.9
6	2.7
8	4.3
10	6.0
12	7.7

Sorption of molecules into zeolites from the gas phase

Some theoretical considerations

Isotherms

The amount of a guest molecule take up into a host zeolite depends upon the equilibrium pressure, the temperature, the nature of the guest molecule and the complexity of the zeolite structure. A common way of analysing some of these parameters is to plot the amount of guest sorbed as a function of pressure at a fixed temperature. This produces a sorption isotherm which can be repeated at different temperatures to compare molecular capacities of a zeolite over the temperature range studied.

An example is shown in Fig. 67 for argon uptake onto zeolite X which illustrates two common isotherm uses in that it shows (i) the variation in gas

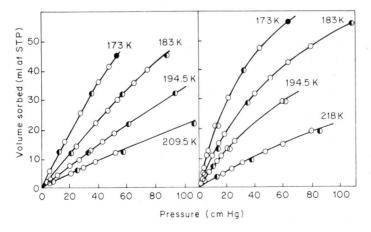

Fig. 67. Isotherm for the sorption of argon into zeolites NaX (left) and BaX (right); calculated ◑, observed ○. Reproduced by permission of the Society of Chemical Industry from L. V. C. Rees, *Chemistry and Industry*, April 1984 (No. 7), p. 253

capacity with pressure and (ii) the influence of ion exchange (Na^+ $= 1/2Ba^{2+}$ in this case). The effect of introducing Ba^{2+} is to increase argon capacity (at 173 and 183 K). This can be related to the change in internal zeolite volume, caused by replacing two Na^+ by one Ba^{2+}. Another use of gas sorption isotherms is to make comparisons of the relative amounts of differing molecules which can be accommodated in the same zeolite. This is demonstrated in Fig. 68 which is an example of the fundamental data from which industrial-scale separations can be designed.

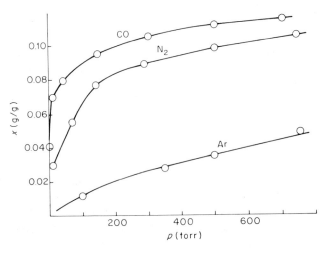

Fig. 68. Sorption of gases into zeolite NaA at 195 K; Uptake x plotted against pressure (p). From D. W. Breck, *Zeolite Molecular Sieves*, Wiley, New York, 1974

The theoretical interpretation of zeolite isotherms has intrigued surface scientists from the early days of gas sorption studies by McBain, who first coined the phrase 'molecular sieve'. Detailed consideration of their interest is beyond the scope of this book but a brief resumé will be attempted.

A common approach has been that of Langmuir wherein the assumption is made that the surface upon which the sorbed molecule sits is one of uniform energy. This is not the case for zeolitic sorption, as quite large variations in sorption site energies arise. This variance can arise in many ways, e.g. from the inherent zeolite geometries whereby molecules are accommodated inside channels and cavities of various shape and sizes. Other factors causing energy heterogeneity are the cations present, silanol groups and the occurrence of water firmly bonded within the framework. Despite the complexity of these factors modifications can be made to the Langmuir equation, used to plot isotherms, to compensate for a proportion of these fluctuations in surface energy.

Another interpretation can be made by likening the sorbate to a liquid phase. Quite good fits can be observed when the density of the sorbed phase is below its critical value, based upon the Hirschfelder relationship:

$$(p+a/v^2+a^1/v^3)(v-b+b^1/v)=RT \tag{7.1}$$

where a, a^1, b and b^1 are coefficients and v is the molar volume of sorbate at pressure p. For fluids of density above their critical value applications of the virial equation become necessary to relate the equilibrium pressure of the sorbed phase to the concentration of sorbate housed in the zeolite cavities.

To conclude this short discussion a comment on the often quoted zeolite surface areas is needed. In other microporous sorbents the well-known BET relationship can be used to calculate available surface areas. This is based upon a multilayer coverage of molecules on the surface and as such is not truly applicable to zeolites, where monolayer coverage usually occurs in the internal surface (i.e. in the channels and cavities)—hence the use of the Langmuir equation may be of more help. A good 'rule of thumb' is that the external surface area of a zeolite is about 1% of the internal surface of the cavities and channels. Note that the sorbate used in the BET method invariably will not gain access to all regions of the zeolite framework.

Isobars and isosteres

When the amount of sorbate taken up by a zeolite at constant pressure plotted as a function of temperature is measured, an isobar is recorded, which is useful in demonstrating the drying capabilities of the zeolites as can be seen in Fig. 69.

Another way of illustrating desiccant properties is via dewpoint plots, like those in Fig. 70, which demonstrate the prevention of 'freeze up' in the cold section of natural gas processing plants. Dewpoints as low as $-100°C$ are readily achieved by using the sodium form of zeolite A for this purpose. Dewpoint plots are a type of isostere which more formally records changes in pressure (p) with temperature when the amount of sorbate taken up by a zeolite is constant. If the data is expressed as a plot of log p versus $1/T$, where T is the absolute temperature (K), the isosteric heat of sorption (q_{st}) can be calculated using the Clausius–Clapeyron equation:

$$q_{st}=4.58\log(1/T_1-1/T_2)\log p_1/p_2 \tag{7.2}$$

This leads to the estimation of a differential heat of sorption which can be used, in turn, to define the more detailed contributions to zeolite–sorbate interaction energies from the factors mentioned earlier, i.e. zeolite geometries, cations, framework charge, etc.

Table XXIII gives some values of q_{st} measured at low surface coverage. The information in this table can be used to demonstrate some of the more obvious variations in an intrinsically complex set of parameters influencing sorption processes. An example of this is seen for CO_2 where q_{st} is seen to increase with increasing cation density (q_{st} NaY < NaX < NaA). The effect of cation charge is clear (q_{st} for Ar in CaX > q_{st} for Ar in NaX), as well as that of hydrocarbon chain length (q_{st} C_2H_6 < C_3H_8 < C_4H_{10} in NaCaA). That dipolar forces make

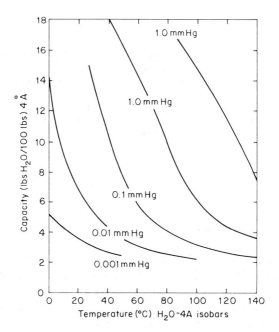

Fig. 69.   Equilibrium isobars for water uptake onto zeolite NaA. From D. M. Ruthven, *Principles of Adsorption and Adsorption Processes*, Wiley, New York, 1984

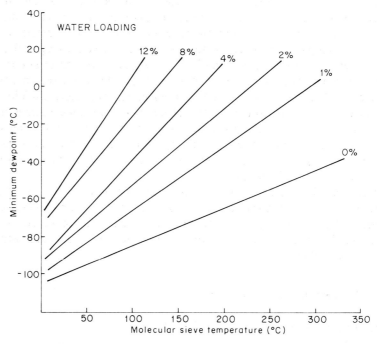

Fig. 70.   Isosteres plotted for zeolite NaA. Minimum attainable dewpoints as a function of water loading and zeolite bed temperature. Laporte Inorganics, Widnes, UK

Table XXIII. Isosteric heats of sorption (q_{st}) measured at low surface coverage for some synthetic zeolites

Sorbate	q_{st} (kJ mol^{-1})				
	NaX	CaX	NaY	NaA	(Na Ca)A
Ar	11.7	20.9	—	11.7	—
CO_2	51	—	34.4	53.5	—
NH_3	75	—	—	—	105
CH_4	18	—	18.1	—	—
C_2H_6	30.5	—	25.8	—	25
C_3H_8	—	—	34.0	—	35
C_4H_{10}	55.6	—	—	—	44

important contributions to the sorbate/substrate interactions is shown by the layer q_{st} values of NH_3, which has a large permanent dipole, compound with those of CO_2, a molecule with a permanent quadrupole moment. In turn CO_2 has a higher q_{st} than that of a spherical sorbate molecule like Ar. (All values were measured in the same zeolite framework, i.e. NaX).

Zeolite capacities

Sorption data, supported by information from NMR, XRD and neutron diffraction for instance, provides quite detailed information as to the amount of sorbates that can be taken up by those zeolites with well-characterized structures. For example argon, nitrogen and oxygen sorption data suggest that these species occupy 755 Å^3 per unit cell in NaA, which closely corresponds to the volume of the α cages calculated from crystallographic data. Water, however, occupies a volume of 833 Å^3 per unit cell approaching the calculated filling of all void space in the A structure (926 Å^3). This implies that some water molecules are in the β cages, in line with other experimental evidence. Of course this information also demonstrates that nitrogen does not easily pass into the β cages, providing one reason why nitrogen surface are measurements are difficult to reconcile with zeolite sorption.

Similar calculations can be applied to other molecules taken up by NaX and Table XXIV lists the maximum capacities attained for some representative molecules. Note that in this list only water penetrates into the β cages (as in A), thus filling the 'total void' available which is close to 50% of the zeolite structure in faujasitic zeolites.

Molecular sieve properties of zeolites

Effect of molecular size

The foregoing paragraph serves as an excellent example of how the unique geometries contained in zeolitic channels and cavities create selective sorption

Table XXIV. Maximum capacity of NaX (Si:Al = 1.25) for various guest molecules

Molecule	No. of molecules per unit cell
H_2O	265
CO_2	120
Ar	140
Kr	116
Xe	74
O_2	149
N_2	134
n-Pentane	34
Neopentane	29
2,2,4-Trimethylpentane	22
Benzene	45
$(C_4H_9)_3N$	16

properties. Not all oxygen windows (rings) are large enough to permit molecular progress through them and the smallest window creating significant separations is the S6R which is, of course, that responsible for the selective sorption of water into the β cages of the A and X structures.

The oxygen windows having the greatest influence on sieving are those composed of 8,10 and 12 oxygens, i.e. the S8R, S10R and S12R rings. Table XXV contains some examples of this influence in zeolites of commerical origin.

However, careful interpretation of Table XXV shows that the ability to separate molecules is not solely a function of oxygen window size. This can be

Table XXV. Approximate correlation between molecular size and effective zeolite molecular sieve aperture

Approx. size (Å)	Example[a]	Window (no. of oxygens)	Molecules accepted
< 3.8	CaM, BaM	8, 12	He, Ne, Ar, CO, H_2O_2 N_2, NH_3, H_2O
< 4.0	NaM, NaA	8	Kr, CH_4, C_2H_6, CH_3OH, CH_3Cl, CO_2, C_2H_2, CS_2, CH_3NH_2
< 4.9	CaA, CaC	8	C_3H_8, n-C_4H_{10}, C_2H_5Cl, C_2H_5OH, $C_2H_5NH_2$, CHF_2Cl, CHF_3, CH_3I, B_2H_6
< 7.8	NaX	12	SF_6, i-C_4H_{10}, i-C_5H_{12}, $CHCl_3$, $(CH_3)_2$ $CHOH$, n-C_3F_8, B_5H_9, CCl_4, $C_2F_2Cl_4$, C_6H_6, $C_6H_5CH_3$, cyclopentane, cyclohexane, pyridine, dioxane, naphthalene, quinoline
< 10	CaX	12	1,3,5-Triethylbenzene

[a] A = Zeolite, A, C = chabazite, M = mordenite, X = zeolite X frameworks, respectively.

seen from the fact that calcium- and barium-exchanged forms of mordenite have a smaller effective window size than that demonstrated by the sodium-exchanged form. A similar phenomenon exists for the A and X zeolites in that, again, Na- and Ca-exchanged varieties have different molecular sieving properties. However, the phenomenon is not the same, in that whereas the introduction of Ca or Ba into mordenite *reduces* the effective window size, the same exchange in both A and X *increases* the apparent window. This will be considered again later.

A further important point is that patently the size limits quoted are *not* those predicted by the crystallographic determination of oxygen ring dimensions. For example, the calcium form of X is quoted as about 10 Å compared to the faujasite structure S12R, measured crystallographically as 7.8 Å. The original names given to the synthetic faujasites by the Union Carbide Corporation were 10X and 13X, for the Ca- and Na-exchanged forms respectively, and were intended to represent their *effective* molecular sieve size in Å—dimensions subsequently shown to be optimistic. The largest molecule accepted into the faujasite structure is decahydrochrysene and the molecule quoted as *not* being taken up is (n-$C_3F_9)_3N$.

The other series of commercial molecular sieves produced by Union Carbide were the 'Linde Molecular Sieves 3A, 4A and 5A'—again the number suggests the pore dimensions and the difference between the sieves is in the cation present. The sodium form (as synthesized) is 4A which can be modified by introducing the potassium cation (3A) or the calcium cation (5A), via ion exchange. Again cations have a specific role in controlling effective pore dimension. Before leaving the subject of oxygen window size, it should be pointed out that the complexities of zeolite framework geometry mean that, in many cases, the oxygen windows are not planar. This is illustrated in Fig. 71 which also shows how ring puckering can alter window dimensions.

Effect of cation exchange

Zeolite A. When water is removed from NaA(4A) those sodium ions in site I (see Chapter 3) move towards the centre of the α cage by about 0.4 Å, but, more importantly, the three sodium ions adjacent to the S8R windows move towards the ring and obstruct the entrance of molecules into the α cage. This results in the creation of an effective 'pore' size of 4 Å. Dehydration of the K^+-exchanged form (3A) places the larger K^+ ion in similar positions, so reducing the pore size to 3 Å. This has been described as the 'sentinel effect' and is illustrated in Fig. 72. When $4Ca^{2+}$ ions are exchanged for $8Na^+$ ions this produces the sieve described as '5A'. This implied dimension arises from a fully 'open' S8R window as the eight ions per pseudo unit cell ($4Ca^{2+} + 4Na^+$) site themselves at site I when water is removed so leaving all site I positions vacant. The effective pore diameter (5 Å) is now close to the crystallographic dimension of the planar eight-oxygen ring of the A structure—i.e. 4.8 Å.

Mordenite. A simple model for the effects of cation siting upon the effective aperture size in mordenite is that based upon the two major channels composing

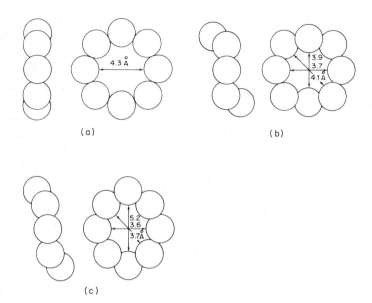

Fig. 71. Diagram illustrating the variation in sizes noted in S8Rs in various zeolite frameworks: (a) A, (b) chabazite and (c) erionite

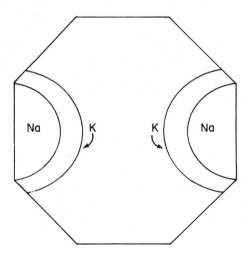

Fig. 72. The 'sentinel' effect created by variation in cation size. Illustrated by the $Na^+ \rightleftharpoons K^+$ sited adjacent to the S8R in dehydrated zeolite A. (NaA = 4A, KA = 3A.)

the structure (see Chapter 3). One of the channels is bound by a series of S8R windows and the other by S12R windows. The differences in sieve action commented on previously can be explained by dehydration, causing the sodium ions to occupy lattice sites which restrict molecular progress down the eight ring channels—hence the effective pore size is that controlled by a puckered S12R. When divalent ions (Ca^{2+}, Ba^{2+}) are introduced they occupy sites in the larger channels (i.e. those proscribed by the S12R windows), so molecular penetration into the mordenite framework is that down S8R channels. When this simple argument is reconsidered, however, it can be seen from Table XXI that the porous nature expected from that controlled by a fully open S12R (i.e. ~ 8 Å) is not realized. The figures quoted relate to a natural mordenite and this has been described as a 'small port' mordenite. The apparent channel blocking has been ascribed both to the presence of extraneous material in the channel structure and to the occurrence of stacking faults—i.e. the framework is not one of continuous channels but one in which the linkage creating the channels have been periodically disrupted.

Synthetic mordenite behaves as a 'large port' sorbent and accepts molecules with critical diameters above 5 Å, i.e. it is closer to the expected S12R behaviour. The channels in natural mordenite can be enlarged by acid leaching and this supports the channel blockage theory.

Zeolites X and Y. An important commercial application of the synthetic faujasites (X and Y) is in the separation of C_8 aromatics. This is based upon ion-exchanged forms containing, say, Na^+, K^+ and Ba^{2+}. Their useful compositions have been arrived at from emperical approaches and, to date, there is no comprehensive explanation of their mode of action. It may be that specific cation distributions between the heteroenergetic sites in the lattice provide the specificity of sorption evident in the separations achieved but it seems unlikely that this is solely a blocking or sentinel effect. In very general terms the cations can be thought of as modifying the anionic framework charge to create a specific internal surface with special affinity for the C_8 isomer to be retained, or held back, from an isomeric mixture.

Other zeolites. Natural clinoptilolite, chabazite and mordenite as well as zeolites L and ZSM-5 have been used as commercial sorbents but no information has been recorded as to the effects that their cation sitings and composition have upon their selectivity. Preparation of clinoptilolite, mordenite (and zeolite Y) in their hydrogen ion form increases sorptive capacity. This can be imagined as arising from the substitution of the 'small' ion H^+ for the larger alkali and alkaline earth cations initially present.

Effect of temperature

Earlier in this chapter the discrepancy between crystallographically measured window sizes and effective pore diameters was pointed out. This anomaly arises from two sources: firstly, both guest molecule and host lattice are not rigid in that both the molecule and the oxygen framework are polarizable (i.e. capable of distortion) and, secondly, both guest and host are in a continuous state of vibration as the bonds holding them together flex under the influence of temperature. This movement creates changes of ~ 0.4 Å in the oxygen window sizes. The net effect of these two phenomena is that molecules of size apparently larger than the measured crystallographic dimensions of the windows can pass through them into the zeolite cavities and channels—hence the concept of 'effective pore size', which is defined by the largest molecule able to pass through the constriction. The size of the molecule is its kinetic diameter which is itself an effective molecular dimension rather than one governed by bond lengths.

A further way in which temperature is influential in molecular sieve separation is as a consequence of the nature of the rate-controlling step which limits sorbate uptake into zeolite crystals. This step is normally diffusion controlled, like that for ion exchange (Chapter 6), and the Arhennius equation is again applicable so that

$$D = D_0 \, e^{-E_a / RT} \tag{7.3}$$

where D is the sorbate diffusion coefficient (i.e. rate of diffusion), D_0 is the pre-exponential factor and E_a the energy barrier presented to the sorbent moving in the sorbate at $T(K)$.

In practice this means that molecules of similar kinetic diameters can be separated on the basis of their time of residence in a zeolite framework. Perhaps the best illustration of this is the use of commercial zeolites A and X to separate the permanent and inert gases from air.

Effect of other parameters

The zeolite literature contains many methods of modification aimed at 'fine tuning' molecular separations on a size basis, but they have not found wide use. Worthy of mentioning are (i) the use of a presorbed polar molecule (e.g. H_2O, NH_3) to create O_2/N_2 selectivity in NaA at 77 K and (ii) the use of silanation to block pores by reacting SiH_4, for example, with OH groups present inside the cavities and channels.

Steaming, which is a common regeneration process for instance when zeolite A is used to dry and 'sweeten' natural gas, causes hydrothermal damage which is described as 'pore closure' and unit cell shrinkage. This can be used to advantage to create useful separations. Recently this effect has been related to aluminium migration from tetrahedral framework sites to create octahedral species in extra framework sites (as shown by MASNMR—see earlier) and also to the creation of a surface skin on zeolite crystals. This skin has important consequences in the separation kinetics in process industries and can arise in the final calcination stage used in the manufacture of zeolite A.

Gaseous diffusion in zeolites

A comprehensive knowledge of the modes of migration of sorbate molecules within zeolite crystals is an important part of industrial applications of zeolites as drying agents, selective sorbents and catalysts. Many fundamental studies have been made to determine the kinetics of molecular entry and subsequent diffusion through zeolite micropores (in both synthetic and natural materials). The rates observed are normally governed by intercrystalline diffusion routes and are temperature dependent as described earlier (see equation 7.3). In some cases, however, industrial studies are carried out on pelletized zeolites. Under these circumstances the rate of progress of molecular species through the macroporous system created by binding the as-synthesized zeolite crystallites into the pelletized aggregate can become important. This means that transport between crystals (intracrystalline) needs to be considered. A schematic representation of the microporous and macroporous cases is shown in Fig. 73.

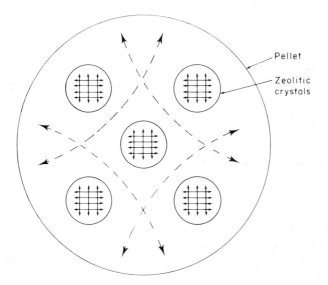

Fig. 73.   Schematic representation of inter- and intra-crystalline diffusion pathways in a pelletized zeolite; intracrystalline ⟨———⟩, intercrystalline) ⟨- - - - -⟩

Methods used to follow the kinetics of sorbate transport in crystals or pellets experimentally can be divided into two groups. One group of techniques follows the transfer of sorbate molecules from the vapour phase to the solid phase (or the reverse). Examples of this are: (i) the use of gas sorption balances, (ii) temperature-programmed desorption (i.e. thermogravimetric methods) and (iii) isotopic tracer measurements. The alternative group of methods are those which measure molecular mobilities within the zeolite cavities or channels. Examples of this group are NMR and IR spectroscopy, neutron scattering and dielectric relaxation techniques.

Sorption and desorption methods

Rates of sorption or desorption can be determined at constant pressure or at constant volume. The fundamental rate law governing diffusion is that of Fick

100

where

$$J = -D(c)\delta c/\delta x \qquad (7.4)$$

where J is the molecular flux, D the diffusivity and c the concentration. The symbol x represents the coordinate along which diffusion takes place, i.e. it is a function of the geometry of the medium in which diffusion is taking place.

The Fickian equation can be solved for the conditions appropriate to the geometry of the system being studied, taking into account the volume and shape of the solid phase in contact with the specific volumes of the gas phase. An example of a convenient equation for use under constant pressure conditions is:

$$M_t/M_\infty = 2A/V(Dt/\pi)^{1/2} \qquad (7.5)$$

where M_t, M_∞ are the amounts of sorbate taken up, or released, at time intervals $t = t$ and $t = \infty$. A and V are the external crystal surface areas and volume respectively and D is the diffusion coefficient. A plot of the fractional attainment of equilibrium (M_t/M_∞) versus $t^{1/2}$ should give a straight line from which diffusional rates (D) can be calculated. Some examples of 'root t' plots, as they are often called, are given in Fig. 74. It can be seen from these that the relationship in equation (7.5) is linear only when t is small. More sophisticated solutions to

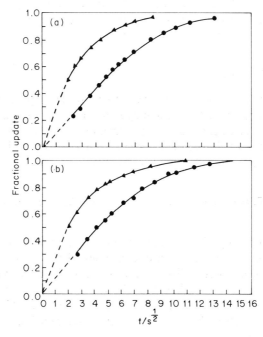

Fig. 74. n-Butane sorption onto zeolite 5A (NaCaA) as a function of root time $(t^{1/2})$ (a) at 310 K and (b) at 323 K and with particle radius 55 μm ($\bullet$) and 27.5 μm ($\blacktriangle$). Reproduced by permission of the Royal Society of Chemistry from H. Yucel and D. M. Ruthven, *J. Chem. Soc., Faraday Trans* 1, **76**, 76 (1980)

Ficks law can be derived to fit diffusional processes occurring over longer time spans.

A similar solution to that in equation (7.5) can be derived for measurements made at constant volume and variable pressure namely:

$$M_t/M_\infty = 2A/V(1 + K/K) \cdot (Dt/\pi)^{1/2} \tag{7.6}$$

The parameter K is a Henry's law constant which is the ratio of the amount of sorbate in the gas phase to that in the solid phase at equilibrium ($t = \infty$).

When experiments, either at constant volume or constant pressure, are carried out at a series of isothermal conditions the variations of D with $T(K)$ can be determined and use of the Arhennius equation (7.3) gives a value of the energy barrier to diffusion (E_a) experienced by the sorbate.

Results obtained from the methods described both assume that the only resistance to mass transport is that of the zeolite structure to the diffusing sorbate. This is not always true; and when small particles are being studied ($\sim 1\,\mu m$ diameter) the external heat generated when sorbent and sorbate are brought into contact with each other can critically affect diffusional rates. Mass transfer resistance can also cause problems and the existence of surface barriers on zeolite crystals has already been mentioned and will obviously play an important part in determining sorption and desorption kinetics. Corrections can be made to compensate for both heat and surface effects to provide a basis of comparison between results measured by sorption techniques to those made by other methods—especially those by NMR.

Use of NMR techniques

Here relaxation times of sorbates perturbed by the appropriate resonance frequencies can be correlated to the average time taken for the agitated molecule to 'jump' between energetic sites with a zeolite structure. The most reliable methods use a pulsed field gradient and recent work is providing an extensive list of diffusion coefficients measured in this way.

Comparison of diffusivities measured by NMR with those obtained from sorption/ desorption methods

To create a valid comparison a correction must be applied to rates measured by sorption as these are concentration dependent. They represent guest molecules moving into an empty void or leaving a filled void to move into 'free' space—clearly a different situation to the NMR where the mobility observed is that of one molecule moving to an adjacent position inside a filled framework.

The correction involves the use of the Darken relationship:

$$\bar{D} = D\delta \ln a/\delta \ln c \tag{7.7}$$

where $\bar{D}$ is the 'differential' diffusivity obtained by sorption and a is the activity of the diffusing molecule at a concentration c. Note that $\bar{D}$ is derived from the mean, or integral, diffusion coefficient, D, which is that measured experimentally over the range $M_\infty \to M_0$ (M_0 is M when $t = 0$). When D is a fraction of concentration the following applies:

$$\tilde{D} = 1/c \int_0^c \bar{D}(c) \, dc \qquad (7.8)$$

When the Darken corrections have been made, D values from NMR are consistently orders of magnitude greater than those measured in similar sorbate/sorbent systems by sorption.

This discrepancy still remains to be fully resolved. The possible presence of surface skins can clearly influence sorption measurements. It should be stressed that these are the conditions of investigations which are closest to the industrial application whereby gaseous mixtures are separated. More detailed studies on the effects of calcination, ion exchange and temperature coupled with structural studies will be needed before this inconsistency can be resolved.

Industrial uses of zeolites as desiccants and specific sorbents

Preparation of zeolite composites

Because synthetic commercial methods produce zeolite crystals which are too small for column use, methods of compacting zeolities into beads or pellets of more appropriate size have been developed. Most industrial processes still use composites in which crystallites are bound together with a clay minerals such as sepiolite, attapulgite or kaolinite. A typical manufacturing process is shown schematically in Fig. 75. The composites can be extracted, or produced as spheres, and will contain 10–20% of clay binder. The presence of the binder can cause problems—especially that of reduction in uptake by 'blinding' the zeolite pores (hence causing poor kinetics). Other problems arise from the creation of catalytic activity causing coke formation and a degraded performance. Problems like these can be overcome and in some cases the use of a binder can be an advantage.

Some products are available without a binder ('binderless'). Their production is like that shown in outline in Fig. 75, except that the zeolite is preformed using kaolin. This intermediate product is then calcined to convert kaolin to meta-kaolin which when subsequently treated with sodium hydroxide creates more zeolite, thus linking existing crystallites into an aggregate of usable attrition resistance and general stability suitable for industrial use.

Ion exchange can be carried out on the formed aggregate as required. This is very useful, particularly when the proposed industrial use requires the introduction of a cation which lowers zeolite thermal stability so that it is damaged by the final calcination stage of the process.

The use of zeolites as desiccants and molecular sieves is not restricted to synthetic species. In countries where natural species are available in suitable

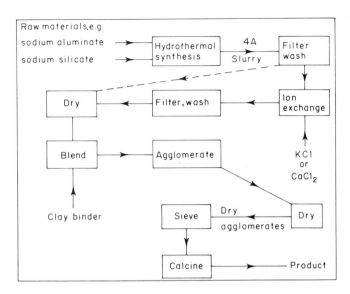

Fig. 75. Diagram representing a typical molecular sieve zeolite manufacturing process. Reproduced by permission of Speciality Chemicals Journal from C. W. Roberts, *Speciality Chemicals*, Feb. 1981, p. 2

quantities and of appropriate purity they are used as mined after dressing and screening to a suitable mesh size.

Use as drying agents

Probably every chemical laboratory in the world has a bottle of 4A beads for use as a general-purpose drying agent, for example in gas drying columns for gas chromatography or for drying an air line.

This illustrates the simplest example of the zeolite affinity for polar sorbents. Their efficiency and capacity for water scavenging is such that in several commercial uses previously dehydrated (activated) composites are sealed into a component which it is required to keep water free during the whole of its working life. Sealed cartridges of 3A, 4A and clinoptilolite are included in refrigeration circuits to dry the halocarbon refrigerants, in vehicle braking and air conditions systems and in heavy duty transformers filled with hydrocarbon liquids. A more recent application of this static desiccant requirement comes in the use of 3A, 4A and 13X in sealed double glazing units containing gases (Ar, SF_6, halocarbons) to improve heat and sound insulation. Zeolites 3A and 4A are preferred because they exclude the filling gas, hence prolonging the window life, unlike other molecular sieves that slowly take in the gases. Zeolites also scavenge out residual vapour from the sealants used to fix the panes together.

On a larger scale zeolite pellets or beads are used to take water from important feedstocks (e.g. hydrogen or oxygen). These are dynamic operations and involve a continuous regeneration regime. Usually reactivation is by a heating– cooling cycle (temperature swing), although pressure reduction and return is also common (pressure swing). Sometimes inert gas purges are used alone or in combinations with temperature or pressure changes.

Both 3A (note that this usually is a partially exchanged NaKA) and 4A (NaA) are used extensively worldwide to dynamically dry liquid hydrocarbons— especially liquid propane gas (LPG), halocarbons and natural gas. 3A is preferred for drying cracked gas (refinery streams containing olefins), ethylene, propylene and methanol (which enters the 4A structure). 4A is chosen for drying other alcohols, benzene and natural gas when 'peak shaving' is required. In the 'peak shaving' requirement the calorific value of the natural gas is adjusted by removing both water and CO_2. The removal of CO_2 using 4A is also important in nitrogen rejection plants.

Use for gas purification

Uses of zeolites (natural and synthetic) to remove impurities from gas streams are too numerous to cover in detail, so only the most well known will be cited.

Perhaps the best known of all purification uses is that of 4A in natural gas treatment which includes the advantage of H_2S removal (i.e. a 'sweetening' process). A similar benefit arises in LPG treatment.

When undesirable larger sulphur- and nitrogen-containing molecules are to be removed from gaseous environments 13X is used (although 5A will remove methylmercaptan). Other general pollution control uses include the trapping of mercury vapour, sulphuric dioxide and nitrogen oxides (NOX). Mordenite (and probably clinoptilolite) can be used when the gas stream has an acidic content, i.e. to remove SO_2 or NO_2 and NO from the respective off gases and HCl from chlorine, chlorinated hydrocarbons and reformer catalyst gas streams (regeneration gas and recycled hydrogen).

Use in drying and purification plants

A typical arrangement for all uses of a drying and 'clean-up' type will use 'twinned' columns of molecular sieves, one in use and one being regenerated at any one time (Fig. 76). The lifetime of a column under normal dynamic conditions is between two and seven years depending upon the operating parameters and gas streams.

Use in bulk separations

The best known separation, based upon a molecular sieving action, is that whereby n-paraffins are accepted by a 5A zeolite whilst i-paraffins are excluded by virtue of their larger effective kinetic diameter. Commercial processes for this separation to provide feedstocks for the detergent and chemicals industry have been developed by Union Carbide (Isosiv), BP, Shell, Texaco (TSF), Exxon

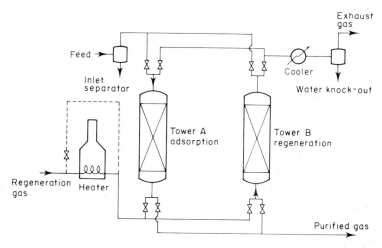

Fig. 76.   Typical layout of a molecular sieve sorption plant. Reproduced by permission of Speciality Chemicals Journal from C. W. Roberts, *Speciality Chemicals*, Feb. 1981, p. 3

(Ensorb) and VEB Leuna Werke (in the GDR). All use 5A (Na CaA) and are gas-phase processes which recover C_{10}-C_{18} linear hydrocarbons from mixtures with branched and cyclic hydrocarbons. In passing it can be remarked that the VEB process also has used a magnesium-exchanged A zeolite. Another large-scale sieving separation is that of the enrichment gases of the air (oxygen, nitrogen and the inert gases). This can be carried out cryogenically but increasingly pressure swing processes at near ambient temperature are being used to generate oxygen and nitrogen for industrial plants.

In Japan and Eastern Europe natural clinoptilolites and mordenites have been used with success in portable pressure swing oxygen generators for use in hospitals or as home oxygen supplies for patients with respiratory problems. Welding and water-oxygenating kits have also been commercially viable based upon the same technology. In the UK both A and X zeolites have been used to generate oxygen for hospital use.

Use in liquid-phase separation

UOP have pioneered the use of synthetic zeolites to fractionate mixtures of liquids. They have developed a series of applications, generically described as 'Sorbex' processes, which operate using a simple tower of sorbent at moderate temperatures and pressures. The tower is divided into zones fed by a rotary valve assembly which controls the flows of feedstock, a desorbent and the product and raffinate streams. The net engineering simulates a moving bed system in which the solid sorbent moves in a countercurrent manner with respect to the liquid phase. Examples of current commercially viable systems of a Sorbex nature are listed in Table XXVI. In 1983 there were 60 units based upon these either in

Table XXVI. Sorbex processes (UOP) for liquid-phase separations

Process	Separation	Zeolite	Desorbent
Parex	p-Xylene from its isomers	KBaY SrBaX KBaX	Toluene p-Diethyl benzene
Ebex	Ethylbenzene from its isomers	NaY or SrKX	Toluene
Molex	n-Paraffins from isoparaffins, naphthalenes and aromatics	5A	Light paraffins
Olex	Linear long-chain paraffins from other paraffins	CaX? SrX?	—
Sarex	Fructose from sucrose and other sugars (corn syrup feed)	CaY	Aqueous system
Sorbutene	But-l-ene from C_4 olefins and paraffins	—	—

operation or at the design stage. Only the 'Molex' process is a molecular size separation and the others function on the basis of a liquid chromatographic fractionation, where the isolation of components from a mixture seems to depend upon very specific interactions between each sorbent molecule and the electro-static fields induced within intracrystalline cavities and channels by the introduc-tion of specific cations combined with a choice of Si/Al (i.e. use of X or Y).

The future

If the number of patents appearing is any indication of likely future applic-ations of zeolites then it seems very probable that their use in pollution control and bulk separation will escalate.

New zeolites, such as L and Theta, with the so-called undimensional channel network, may become important, whilst the ability of Y to accomplish separ-ations from aqueous environments will undoubtedly be extended. Finally the increased commercial availability of high-silica zeolites (e.g. ZSM-5) is already creating great interest in systems where hydrophobic molecules are taken up from hydrophilic ones by the essentially water-repellant surfaces of these zeolites. Particular attention has been given to the potential recovery of alcohol from fermentation broths and in biotechnological applications.

Gas chromatography

The use of zeolites in gas–solid chromatography (GSC) should not be forgotten. The analytical separation and quantification of permanent gases on 5A was one of the earliest GSC determinations. Other useful methods are the detection of butane isomers and other hydrocarbons and the analyses of isotopic mixtures, e.g. those of hydrogen, carried out at $-196°C$.

CHAPTER 8

The stabilities of zeolite structures and their modification

Introduction

It is useful to preface this chapter with a reminder that all zeolites are metastable species by geological definition. The non-occurrence of some synthetic species in nature (e.g. zeolite A) can easily be explained on this basis. Accordingly the fact that zeolites subjected to modest chemical and physical treatments can often be degraded should not come as a surprise. Furthermore it is a useful concept to remember that the image of the aluminosilicate framework as a rigid immutable entity is not confirmed on inspection. Even the oxygens in the framework exchange with gas-phase oxygen molecules under mild conditions and direct substitution for silicon and aluminium in tetrahedral sites is relatively easy—especially in high-silica zeolites.

Thermal stability

Several levels of temperature resistance can be illustrated, e.g. conversion from one zeolite structure to another has been observed even at temperatures close to ambient. This can be demonstrated by reference to synthetic gismondines. The structure described as NaP (also known as B) can be synthesized quickly and easily even at room temperature. It first crystallizes in cubic form (NaP_C) and readily transforms to a tetragonal form (NaP_T); this will eventually take place even at room temperature, but it can be observed by isothermal differential thermal analysis (DTA) within an hour or so at an elevated temperature (Fig. 77). The $NaP_C \rightarrow NaP_T$ transformation is a short timescale version of an order–disorder change which is a fundamental one to all solid-state transformations.

Another commonly observed transformation of a zeolite structure is that promoted by water loss. This can cause a change from one zeolite structure to another, structure collapse to an amorphous phase or recrystallization to non-zeolitic materials. These critical water losses can occur over a wide temperature range. Zeolites of low stability are paulingite and phillipsite, which recrystallize

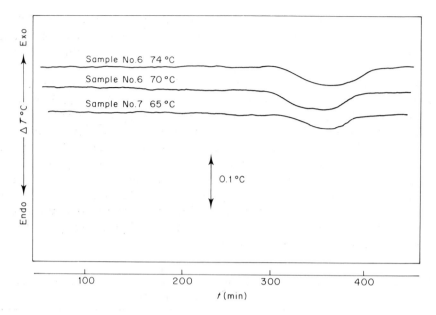

Fig. 77. Isothermal DTA traces showing a thermal event at about 6 h caused by the NaP$_c$→NaP$_T$ transformation

to new phases at about 250°C, and at the other end of the scale (i) analcime, A, X and Y are stable to at least 700°C and (ii) mordenite and offretite retain their structures to about 800°C and 900°C, respectively. These figures are only approximate as cation content can drastically alter temperature stability. An extreme example of this is that of the barium-exchanged forms of zeolite A. This loses structure at temperatures much less than 100°C and can be made to

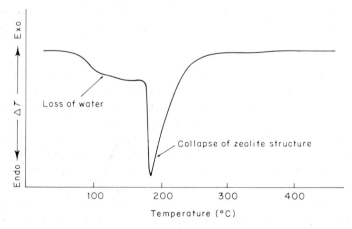

Fig. 78. DTA of zeolite BaA. The first endotherm represents loss of water and the second the onset of lattice collapse

collapse on removing water by vacuo even at ambient temperature (Fig. 78). In other cases the introduction of a multivalent cation, by ion exchange, causes hydroxyl group formation. Dehydroxylation of these groups can alter stabilities——not always deleteriously however.

Anhydrous zeolite frameworks ultimately collapse to form amorphous materials or recrystallize to silica phases (e.g. tridymite and cristobalite), feldspars (e.g. celsian, $BaAl_2Si_2O_8$, and high sanidine, $KAlSi_3O_8$), feldspathoids (e.g. nepheline, $Na_6K_2[(AlO_2)_8(SiO_2)_8]$) or even to clay minerals (e.g. montmorillonite).

A simple classification of common zeolites is shown in Table XXVII on the basis of thermal stability. This should be taken only as a'rough guide and the figures quoted may well be based upon data produced on outmoded thermal analysis equipment. In addition they may be very far from the conditions reflected in industrial usage. Despite these criticisms thermal analysis can be very useful to determine water losses, dehydroxylations and phase changes as it is a very convenient, easy technique to use.

Table XXVII. Approximate classification of some zeolites by their temperature stability as defined by thermal analysis.

Temperature range of stability	Zeolite
Ambient to 250°C	Harmotome, phillipsite, paulingite, heulandite[a]
250 → 400°C	Gismondine, yugawaralite, stilbite, brewsterite, stellerite
400 → 600°C	Laumontite, faujasite, natrolite, scolecite, mesolite, thomsonite, gonnardite, edingtonite, P, ZK-4
Over 600°C	Analcime, erionite, offretite, chabazite, mordenite, clinoptilolite, bikitaite, wairakite, dachiardite, merlionite, mazzite Omega, A, X, Y, L, ZSM-5.

[a] Changes to heulandite B at 250°C.

Some thermal profiles are shown in Figs. 79 and 80 to illustrate the comparative use of thermoanalytical data. Fig. 79 shows how cation content alters water loss events and several authors have persued this to relate them to complex cation–water, water framework and cation–water framework interactions. The effect of cation content on water capacity can be easily seen from thermogravimetry (Fig. 80). Thermomechanical analysis (TMA) can quantify the onset of structural collapse (Fig. 81). All thermoanalytical methods need the support of other methods, especially XRD, to confirm the interpretation of thermal profiles.

Other areas where thermal methods have been usefully employed are (i) the detection and quantification of those dehydroxylation processes (Fig. 82) which have been closely identified with catalytic activity (as will be seen later) and (ii)

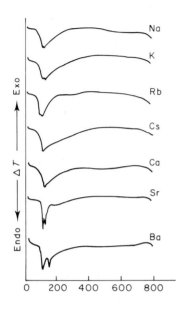

Fig. 79. DTA traces of various homoionic forms of clinoptilolite

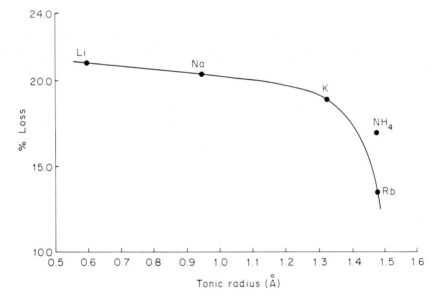

Fig. 80. Percentage water losses plotted against cation radius for ion-exchanged chabazites. (Percentage water loss determined by thermogravimetry.)

the use of temperature-programmed desorption (TPD) to follow the loss of a sorbate from specific zeolite sites.

Although not a measure of zeolite stability it is convenient to expand on the TPD technique at this point. As the name implies the methodology of the TPD

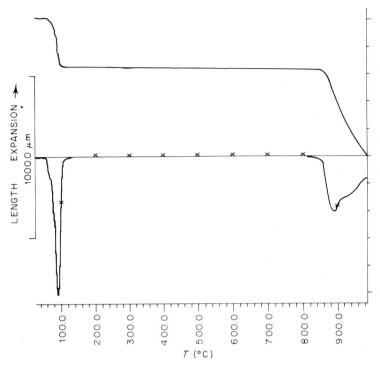

Fig. 81. TMA of Cu-exchanged zeolite L. Top trace represents change in dimension with temperature. Bottom trace is a derivative. The first peak at ~98°C represents unit cell shrinkage due to water loss with subsequent relaxation. The second peak (>800°C) is caused by lattice collapse

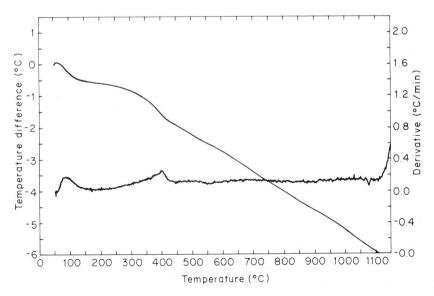

Fig. 82. The DTA (upper line) and derivative (lower line) of a Mg-exchanged zeolite. The peak at 400–450°C represents a dehydroxylation event

approach is to preactivate a zeolite sample and then to allow it to come to equilibrium with a sorbate molecule. Subsequently the sorbate/sorbent system is heated at a controlled heating rate with the resulting sorbate loss followed by thermogravimetry, gas chromatography (GC), mass spectrometry (MS) or by combined techniques, e.g. GC/MS.

The thermal profiles produced can be used to measure accurately the interaction of the sorbate with the sorbent and so reflect the influences of zeolite structure, nature of cations present, lattice charge (Si:Al ratio) and the presence of catalytically active sites. Fig. 83 illustrates a typical TPD profile.

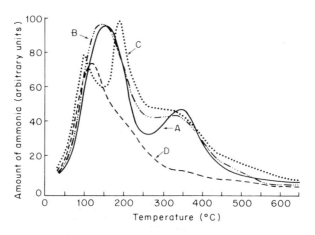

Fig. 83. Typical TPD profiles. The curves are loss of ammonia from ZSM-5 zeolites; A = HZSM-5, B = Ni/HZSM-5, C = MgHZSM-5 and D = PHZSM-5. Desorption peaks correspond to the loss of NH_3 from acid sites which can be seen to vary in strength with the element used to modify the ZSM-5. Reproduced by permission of Elsevier Science Publishers BV from M. Derewinski et al., in P. A. Jacobs et al., (eds.), *Structure and Reactivity of Modified Zeolites*, 1984, p. 213

The recent improvements in the design and sensitivity of thermal analysis, in association with microprocessor capability to analyse profiles in detail, means that this technique is beginning to achieve its full potential as a research and analytical tool for zeolite studies. To illustrate the detail which can be obtained from TPD profiles obtained from a thermogravimetric balance coupled to a dedicated microprocessor, Fig. 84 has been included. This plots the variation in the energy of activation experienced by four sorbate molecules, as a function of the number of molecules present per unit cell (m/uc), as they leave the sorbent—a pentasil zeolite.

Hydrothermal stability

When zeolites are exposed to water vapour at elevated temperatures and/or pressures they are experiencing conditions close to those under which they are

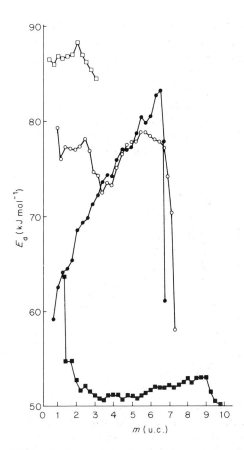

Fig. 84. Variation of E_a for the desorption of sorbates recorded as molecules sorbate per unit cell (m/uc) for a high-silica zeolite; n-hexane ●, n-octane □, ethanol ■, p-xylene ○. Reproduced with permission from R. E. Richards and L. V. C. Rees, *Zeolites.* **6**, 23 (1986). Copyright Butterworth Publishers

formed in nature. Under these circumstances they mimic the natural progression to a more stable and less porous phase. If the experimental timescale is long enough, at extreme conditions many zeolites yield an analcime phase. Studies of hydrothermal stability are of great importance to the regeneration processes used in association with the use of synthetic zeolites as catalysts, molecular sieves and drying agents as mentioned in other chapters.

Stability to acids

It is often assumed that zeolites have a low resistance to mineral acids and indeed both A and X readily dissolve in even a modest molarity of hydrochloric acid. This ease of dissolution can be linked to the ready removal of aluminium from tetrahedral site frameworks where Si: Al is $1 \rightarrow 2$. The leached aluminium readily

114

hydrolyses to a variety of species in which the metal is hexacoordinated, as confirmed by MASNMR.

At higher Si:Al ratios the effect of mineral acid still leaches framework aluminium, but in some instances a zeolitic structure can be retained. Clinoptilolite, mordenite and ferrierite have been shown to retain crystal habit and integrity even after six months exposure to 8M nitric acid (Fig. 85). Leaching can be promoted by agencies other than mineral acids and the same effect can be attained by treatment, for example, with ethylenediamine tetraacetic acid (H_4 EDTA), silicon tetrachloride, fluorosilicates, organic acids and even acetylacetone, which forms a stable Al complex. It has been claimed that clinoptilolite, mordenite and ZSM-5 can be stripped of all aluminium to leave a silica pseudomorph of each structure.

Fig. 85. Scanning electron micrographs of clinoptilolite exposed to nitric acid of increasing molarity: (a) untreated. (b) 1M, (c) 5M and (d) 8M

In fact recent careful studies on the reaction of zeolites with moderate acid molarities demonstrate that the first stage is that of a cation exchange whereby hydronium ions (H_3O^+) replace the indigenous cations. This can be carried out even in zeolites A and X and can be linked to the hydrolysis phenomenon which includes the same exchange process (see Chapter 6).

If the hydronium form of a zeolite is heated, water is lost and a 'hydrogen zeolite' is formed, viz:

$$H_3O^+Z^- \rightarrow H^+Z^- + H_2O$$

When the zeolite Si:Al ratio is high these 'hydrogen zeolites' are stable and are desirable catalysts.

A complication to this phenomenon is that, during cation exchanges involving the introduction of polyvalent transition elements into a zeolite, localized acid conditions prevail, creating the simultaneous hydronium exchanges previously mentioned, with the adjunct possibility of creating acid sites on subsequent calcination.

Yet another way to introduce the H^+ species is via a route described originally as 'decationization' or "deammination". This involves a partial exchange with ammonium ions followed by a careful calcination to decompose the ammonium species. This was first described by McDaniel and Maher in 1968, for zeolite Y, as

$$NaY + NH_{4(aq)}^+ \rightleftharpoons NaNH_4Y + Na_{(aq)}^+$$

$$NaNH_4Y \xrightarrow{\text{calcine}} NaHY + NH_{3(g)}$$

This proved to be a route to a very stable acid-cracking catalyst ('ultrastable Y'), able to resist temperatures to over 1000°C.

Stability to alkalis

The presence of the OH^- species in solution is a critical factor in some zeolite transformations. The pH of the gel precursor in syntheses determines products and so it is to be expected that alkaline treatment of existing crystalline zeolites might create structural changes.

A sample is that of zeolite A which, when exposed to even dilute NaOH, transforms to a gismondine phase (NaP). In higher concentrations hydroxysodalite (HS) readily forms and this was a common hazard in the early days of commercial A syntheses. Prolonged exposure (months) to NaOH ultimately dissolves the A structure.

Similarly, extended contact of clinoptilolite to the highly alkaline 'pond water' (see Chapter 6) ultimately degrades the zeolite, causing attrition and hence loss of column efficiency. Some evidence suggests that faujasite structures are formed during this treatment. To avoid this the SIXEP plant (Sellafield BNF) operates at $\sim$pH 7.

Other chemical modifications

One of the major factors in the importance of zeolites as catalysts is their ability to accommodate a wide variety of chemical modifications with relative ease whilst retaining their open structures.

Obviously ion exchange is the major way to achieve modification but other facile isomorphous replacements can be carried out to introduce heteroatoms into tetrahedral framework sites. This seems especially easy to do in MFI zeolites where even aqueous solution treatment can introduce, for example, B and Ga into tetrahedral positions.

Literature claims, especially those in patents, suggest that any tri-or tetra-valent species can be isomorphously placed into framework positions. In many cases these claims remain to be substantiated. A distinction should be drawn between replacements made via modification of existing zeolites, which are being considered here, and those structures which can be synthesized with atoms like P, Co, Zn and Sn in tetrahedral sites, which are described as zeotypes and are considered in Chapter 10.

Another modification commonly made is to introduce salt molecules into the zeolite pore structures. This can be done by use of molten salts or strong electrolyte solutions and it is possible to fill the zeolite voidage in this way. An important aspect of this lies in the use of 'passivating' agents to improve the performance of zeolite cracking catalysts. This will be mentioned in Chapter 9.

A final example of isomorphous replacement is to make use of the mobility of the oxygen atoms to substitute them by fluorine. Again this can be carried out by molten salt or salt solutions.

Modification by silylating agents can be used to alter zeolite surfaces to create hydrophobic properties or to partially block pores and so alter sieving behaviour. Silylation also provides a route to link chelate groups to zeolite surfaces. This has been used to prepare selective ion exchange materials for use in ruthenium isotope removal from simulated aqueous nuclear wastes.

Stability to ionizing radiation

Zeolite structures have proved to be remarkably resistant to radiation and prolonged exposure to high neutron and γ doses produces a negligible effect on the zeolite matrix. High β doses are equally ineffectual. Clearly these are the properties needed for the use of zeolites to treat intermediate-level aqueous nuclear wastes and their subsequent containment in cement matrices.

CHAPTER 9

Zeolites as catalysts

Introduction

The first use of zeolites as catalysts occurred in 1959 when zeolite Y was used as an isomerization catalyst by Union Carbide. More important was the first use of zeolite X as a cracking catalyst in 1962, based upon earlier work by Plank and Rosinski. They noted that relatively small amounts of zeolites could usefully be incorporated into the then-standard silica/alumina or silica/clay catalysts. The use of zeolites in this way as promoters for petroleum cracking (i.e. the production of petrol from crude oils) greatly improved their performance. The resultant saving from this improvement can be measured in billions of pounds per annum in the present economy. It is this saving that has been responsible for the huge amount of money and time which has been invested in zeolite research during the last 25 years.

A comprehensive account of this research is well beyond the scope of this introductory text so only the major concepts and uses will be described. The subject is infinitely complex and often difficult to relate to industrial reality, due to the understandable reluctance of the oil and chemical companies to furnish details of which zeolites are used or indeed even which industrial processes can be commercially successful when relatively expensive zeolite catalysts are used.

What is clear, however, is that the major employment of zeolites is as acid cracking catalysts. As such they account for over 99% of the world's petrol production from crude oils. A reasonable estimate of the world consumption of zeolites for this purpose is that some 400,000 tons are used annually. Of this some 350,000 tons are used for cracking.

A modern 'cat cracker' will make use of the zeolites in a fluidized bed reactor and Fig. 86 illustrates their worldwide FCC usage in 1980. This, and the other commercially viable uses of zeolites as catalysts, will be described later but all are based upon the very special properties that zeolites possess and it is probably true to say that zeolites can be made to catalyse *any* reaction on the basis of their high surface areas alone. It is also important to say that the usefulness of this versatility is conditioned by economics, but it is appropriate to explain this flexibility in more detail.

117

118

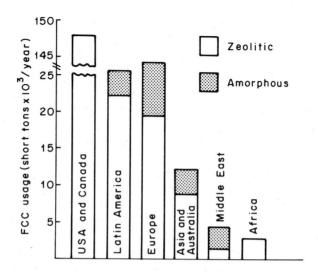

Fig. 86.   Usage of FCC catalysts (1980). Reproduced by permission of the Royal Society of Chemistry from D. E. W. Vaughan in R. P. Townsend (ed.), *Properties and Applications of Zeolites*. Chemical Society Special Publication 33, 1980, p. 298

Potential versatility of zeolites as catalysts

Vaughan has graphically described zeolites as 'molecular boxes' which have variable dimensions suited to the encouragement of molecular rearrangements inside their confined geometry. The conditions inside the 'box', and of the 'box' itself, can be controlled in a variety of ways based upon the unique properties of zeolite frameworks as summarized in Table XXVIII and described below.

TABLE XXVIII.   Correlation between zeolite properties and catalytic functionality

Property	Catalytic functionality
Crystal voidage and channels	Extensive internal surface to encourage catalytic processes
Variable pore size	Creates both reactant *and* product selectivity via molecular sieving
Ion exchange	Cations (i) control pore size, (ii) create high potential energy fields within voidage (active sites) and (iii) enable distribution of catalytically active metals on the zeolite substrate
Salt occlusion	Controls pore size, provides another method of metal incorporation and can improve thermal stability and poisoning resistance.
Framework modification	Varies lattice charge (by synthesis or modification) to enhance active site production and thermal stability

Crystal voidage and channels

Although some heterogeneous reaction will take place at the external crystal surfaces, most practical zeolite catalysis takes place inside the framework. Here zeolites have the advantage of a very large internal surface, about 20 times larger than their external surfaces for the more open frameworks (e.g. zeolites X and Y). This internal capacity provides the appropriate surfaces at which catalytic transformations can take place. In the faujasitic zeolites this is typically in the series of large cavities easily available via three-dimensional open-pore networks.

Further flexibility which is useful for planned catalytic uses arises in the more recently produced zeolites with subtly different cavity and channel systems. ZSM-5, for instance, has a three-dimensional system linked via intersections rather than cavities and mordenite catalysis seems to take place only in the largest channels.

At the time of writing the overwhelming use of zeolite catalysts rests with the synthetic faujasites, mordenite and ZSM-5. All other zeolites await the development and exploitation of their individually unique channel systems.

Variable pore sizes

Given that catalysis take place largely within zeolite frameworks, access to this environment is patently controlled by the oxygen windows. This is a diffusion-limited process, as is the egress of product molecules after transformations have taken place. This means that zeolites have very special practical advantages over the more traditional catalysts, in that they will admit only certain reactant molecules and that this can be potentially tailored to produce selected products. This selectivity is known as 'shape-selective catalysis' and is controlled by 'configurational diffusion'—this phrase was coined by Weiss to express a diffusion regime in which useful catalytic reactions are promoted by virtue of a matching of size, shape and orientation of the reactant and product molecules to the geometry of the zeolitic framework. This will be described in more detail later, but Fig. 87 shows the simple principle behind this reactant and product shape selectivity.

Ion exchange

Chapter 7, describing the molecular sieve action of zeolites, has explained the role of cations in controlling pore size. In a catalytic context this can, in theory, be used to tailor zeolites to achieve special shape selectivities.

Perhaps more relevant is the way in which ion exchange can be employed to place cations into very specific framework sites so as to create small volumes of high electrostatic field. These fields are 'active sites' to which an organic reactant molecule can be attracted thus promoting the bond distortion and rupture essential to molecular rearrangements. The exact nature of these, and other, active sites has been the subject of much scientific thought, as will be seen later.

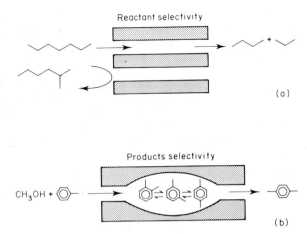

Fig. 87.   Representation of (a) reactant shape selectivity in zeolite channels (rejection of brached chain hydrocarbons) and (b) product shape selectivity (*p*-xylene diffuses preferentially out of the channels). Reproduced by permission of the Royal Society of Chemistry from S. M. Csicsery, *Chemistry in Britain*, May, 1985, p. 473

Another feature of ion exchange is that it provides a route for the introduction of metal cations with a view to their subsequent reduction to metal particles. These exist in the so-called 'bifunctional' zeolite catalysts used to effect both hydrogenation and dehydrogenation reactions.

Salt occlusion

The introduction of a salt molecule into a zeolite can be the first stage in the incorporation of a metal for subsequent reduction as mentioned above. It can also be used to enhance thermal stability. Yet another purpose is to 'pacify' zeolite cracking catalysts. The problem here is that crude oils contain metal cations (Ni, Cu, V, Fe) originating from the metal porphyrins thought to play an inherent part in the geological formation of oil. These metals create unwanted reactivity causing carbon (coke) formation and subsequent loss of catalytic properties. The occlusive introduction of stannates, bismuthates or antimonates pacifies these metals to extend useful catalyst bed life. It enables the refinery to cope with a variety of crude oils from different oil fields and illustrates the flexible technology which can be achieved in zeolite catalysis.

Other salt treatments, via phosphates or fluorides, have been used to improve performance.

Framework modification

The electrostatic field inside a zeolite can be manipulated by isomorphous substitution into framework Si and Al sites. This can be done by synthetic or modification routes. When the Si:Al ratio is close to 1 the field strength is at its highest as is the cation content—i.e. the conditions of maximum negative charge on the framework. An increase of the Si:Al ratio causes a greater separation of negative charge and hence higher field gradients (obviously also conditioned by cation position and cation type). In this way catalytic activity can be controlled, and other parameters altered. A well-known example of these effects is the way in which the thermal and chemical stabilities of the synthetic faujasites can be critically altered by aluminium removal (see Chapter 7).

Framework substitution also can be created by the introduction of atoms other than Si and Al into tetrahedral sites via synthesis or modification. ZSM-5 can accept B and Ga into tetrahedral sites by simple salt treatment as mentioned earlier, although a similar reaction in other frameworks is by no means as facile.

Zeolite active sites

Acid sites

Most industrial applications of zeolites are based upon technology adapted from the acid silica/alumina catalysts originally developed for the cracking reaction.

This means that the activity required is based upon the production of Brønsted sites arising from the creation 'hydroxyls' within the zeolite pore structure. These hydroxyls are usually formed either by ammonium or polyvalent cation exchange followed by a calcination step, viz:

Ammonium ion exchange

$$NaZ_{(s)} + NH_{4(aq)}^+ \rightleftharpoons NH_4Z_{(s)} + Na_{(aq)}^+$$

$$NH_4Z_{(s)} \xrightarrow{calcine} NH_{3(g)} + HZ_{(s)}$$

Polyvalent ion exchange

$$NaZ_{(s)} + M(H_2O)_{(aq)}^{n+} \rightleftharpoons M(H_2O)^{n+}Z_{(s)} + nNa_{(aq)}^+$$

$$M(H_2O)^{n+}Z_{(s)} \xrightarrow{calcine} MOH_{(s)}^{(n-1)} + HZ_{(s)}$$

In high-silica zeolites (e.g. ZSM-5 and clinoptilolite) these 'protonated' zeolites (HZ) can be made by direct exchange with mineral acid. Ideally the 'protonated' form contains hydroxyls which are protons associated with negatively charged framework oxygens linked into alumina tetrahedra, i.e. Brønsted sites are created:

The protons have great mobility when the temperature is above 200 °C, and at 550 °C they are lost as water (clearly seen by thermal analysis) with the consequent formation of Lewis sites:

$$\text{Brønsted site} \longrightarrow \text{Lewis site}$$

The Lewis sites in turn are unstable, especially in the continued presence of water vapour and an annealing process stabilizes the structure. This produces the so-called 'true' Lewis sites by ejecting Al species from the framework, i.e.:

$$\text{Lewis site} \longrightarrow \text{'True' Lewis site}$$

This extra framework aluminium can be identified by MASNMR (see Chapter 4), but is in at least two forms. Theoretical calculations support the idea that $(A10)^+$ species are the source of useful Lewis activity, whilst suggesting that the tricoordinated Al of the Lewis site (Al* above) acts as a weak acid. (There is however *no direct* evidence for the existence of tricoordinated Al.) Electron spin resonance (ESR) studies propose that, when Brønsted sites interact with nearly Lewis sites, 'super acid' sites are formed. The evidence is that these seem likely to be due to displaced aluminium species. The uniqueness of zeolites as cracking catalysts lies in the high density of these active sites coupled with the zeolite inherent stability and amenability to regeneration.

Identification and quantification

Infrared spectroscopy has been responsible for much of the earlier work contributing to the understanding of the mechanisms invoked in zeolite catalysis. It shows that two major hydroxyl species can be identified in zeolite substrates. The first has an absorption band at 3740 cm^{-1} and is diagnostic of terminal hydroxyls of the form:

These are well known from other silicates and silica itself. In a zeolite framework they may arise at the surface or at defect sites in the framework. The second hydroxyl has an absorption band in the 3600–3650 cm^{-1} region and can be assigned to OH groups associated with a Brønsted site. They are known as 'bridging' hydroxyls.

Proton MASNMR confirms these hydroxyl positions as well as other species. These other moieties help to an elucidation of the role played by 'H' species and include Al–OH (non-framework), residual NH_4^+ and other acidic OHs.

Quantification of acid strength can be made by a variety of methods—largely associated with the observation of the influence caused by an sorbate molecule assumed to interact with an active site. Infrared studies measure acidities by observation of bathochromic shifts created by the way in which a sorbate affects the bridged hydroxyls.

This has been clearly illustrated by the correlation with a Sanderson electronegativity, calculated by taking into account zeolite Si:Al ratios and structure, as shown in Fig. 88. This correlation is based upon benzene as a sorbate and the increase in bathochromic shift illustrating the weakening of the OH bond created by increased interaction between benzene and zeolitic hydroxyls. Electronegativity increases with the Si:Al ratio.

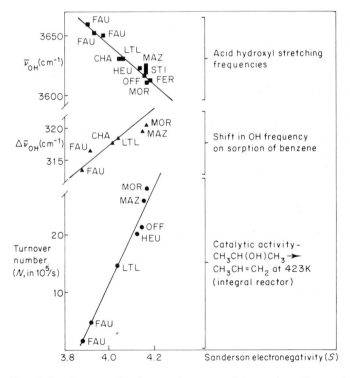

Fig. 88. Correlation between Sanderson electronegativity and zeolite catalytic properties; $\Delta\bar{v}$ = wavelength of OH bond and N = turnover number, an experimental parameter used to quantify catalytic activity. Reproduced by permission of the Society of Chemical Industry from J. Dwyer, *Chemistry and Industry*, April, 1984 (No. 7), p. 261

Other quantitative measures of acidity include direct titration with a Hammett indicator and desorption of sorbed ammonia by TPD. Pyridine sorption can be used to measure the number of acid sites present in a zeolite by (i) infrared spectroscopy, using bands at 1545 and 1450 cm^{-1} assigned to pyridine located on Brønsted and Lewis sites, respectively, or (ii) by the changes noted in proton MASNMR spectra.

Basic sites

In certain instances reactions have been shown to be catalysed at basic (cation) sites in zeolites without any influence from acid sites. The best characterized example of this is that of K-Y which splits n-hexane isomers at 500 °C. The potassium cations have been shown to control the unimolecular cracking (β-scission). Free radical mechanisms also contribute to surface catalytic reactions in these studies.

Shape-selective catalysis

Reactant selectivity

The control of a catalytic process based upon the exclusion of a potential reactant can be simply illustrated for butanol dehydration. Using CaA, n-butanol can enter the zeolite and be converted whereas i-butanol cannot. On the other hand both can enter NaX and both are dehydrated. This is demonstrated in Fig. 89.

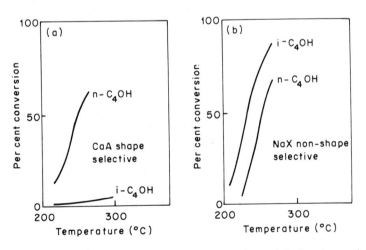

Fig. 89.   Shape-selective dehydration of n-butanols on (a) CaA (only n-butanol converted) and (b) NaX (both n- and i-butanol converted). Reproduced by permission of the Society of Chemical Industry from J. Dwyer, *Chemistry and Industry*, April, 1984 (No. 7), p. 261

Product selectivity

This is essentially a function of the rates of diffusion of products away from the reactive sites, through the crystal pores and out of the crystallites. Perhaps the best known example of this type of selectivity is in the preferred alkylation of toluene to *p*-xylene over H-ZSM-5, because the diffusivity of the *para* isomer, in the zeolite, is much greater than that of the *o*- and *m*-xylene isomers in the same substrate.

Restricted transition state selectivity

Control of product shape can arise because the volume available round an active site in a zeolite framework is such as to preclude the formation of a bulky intermediate product (transition state) of the reactants meeting at the site. This is illustrated in Fig. 90, which shows how the acid-catalysed transalkylation of a dialkylbenzene is thought to proceed in a mordenite substrate. The product (*m*-xylene) passes into the main channels of the mordenite and forms a diaryl intermediate probably via:

Transition state A

or

Transition state B

It can be seen that transition state A cannot 'fit' in the mordenite channels and so this precludes the production of molecules arising from scission of this transition state (i.e. 1,3,5-trimethylbenzene and toluene). On the other hand transition state

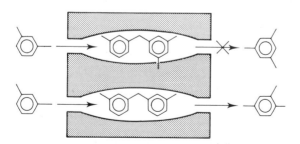

Fig. 90. Transition state shape selectivity in zeolite channels (see text). Reproduced by permission of the Royal Society of Chemistry from S. M. Csicsery, *Chemistry in Britain*, May, 1985, p. 474

B can be accommodated within the channels and so its breakdown products (i.e. 1,2,4-trimethylbenzene and toluene) are the ones which form.

Shape selectivity can be induced by altering other zeolite parameters—for instance crystallite size variation as mentioned earlier in this chapter. Larger crystals provide longer diffusion path lengths and this can be used to modify catalysis products because product selectivity is a function of a diffusion-limited process. An illustration of this is the reduction in the formation of durene from methanol on ZSM-5. This unwanted product can be reduced in its creation by a factor of two by a 10–20 fold increase in zeolite crystallite size. Inherent in this parameter change is the time element, of course, so 'time on stream' can alter product formation based upon their diffusional rates. A neat variation on this is to make use of the normally deleterious formation of carbon deposits during cracking to alter activity and shape selectivity of H-ZSM-5, for instance— presumably by reducing diffusion path lengths by the partial blockage of channels and/or intersections by coke. Modification by P, Mg or B of H-ZSM-5 has been shown to create changes in the products from the alkylation and disproportionation reaction of toluene. These changes in selectivity have also been assigned to alterations in effective diffusivities arising from the intrusion of carbon, or oxides, into the critical channel structures.

Bifunctional catalysis

Part of the modern hydrocracking process, whereby crude oils are cracked in an atmosphere of hydrogen gas, calls for the availability of a catalyst having the ability to function as an aid to both hydrogenation and dehydrogenation processes. This is best attained by the dispersion of a reduced metal on an inert support. The creation of such a catalyst on a zeolite substrate has been the subject of considerable investigation—not the least because they provide model systems in which to examine metallic clusters, i.e. aggregates of relatively small numbers of metal atoms. Metals such as Ag, Pt, Pd, Rh, Ni and Fe have been incorporated into zeolites in several ways. The most frequent has been through an initial cation exchange using the appropriate aqueous halide salt, although complex cations (e.g. $[Pt(NH_3)_4]^{2+}$) and metal carbonyls have often been alternative routes. The sorption of $Mo(CO)_6$, from the vapour phase, onto Pt metal particles dispersed in zeolite Y to create a bimetallic PtMoY catalyst is an example of the carbonyl route.

The reduction step of the metal cation to metal particles usually involves a preliminary careful evacuation at about 400 °C followed by an oxygen gas purge. The final reduction is by hydrogen gas in the 400–600 °C temperature range. Reduction by carbon monoxide is an alternative.

The metal aggregates formed on zeolites by these treatments exist both inside the zeolite cages and as surface species. They have been largely characterized by transmission electron microscopy coupled with XPS studies to quantify surface enrichment. Recently Fraissard has demonstrated a novel NMR method to study metal dispersions inside zeolite cages. This uses the NMR chemical shift of

^{129}Xe which can be related to the number of hydrogen atoms covering each metal aggregate. By calculating the number of hydrogens taken up by a metal-loaded zeolite, together with a knowledge of the metal concentration, an average number of constituent metal atoms in each aggregate can be estimated.

Commercial processes using zeolite catalysts

The major commercial processes making use of zeolite catalysts are listed in Table XXIX. Again it should be stressed that this should be regarded as a guide and largely relates to practice in the Western world. Reviewing zeolite catalyst literature and patents suggests that other areas of the globe may make use of natural clinoptilolites and mordenites in their oil and chemical industries. A more detailed analysis of the processes in Table XXIX will now be made.

TABLE XXIX. Commercial processes using zeolite catalysts

Process	Catalyst	Advantage in using zeolite-based catalysts
Catalytic cracking	REY (REX, REHY, REMgY, HY)	Selectivity and high conversion rates
Hydrocracking	X, Y, mordenite, erionite loaded with Co, Mo, W, Ni, also HY, US-Y, Ca MgY and H-ZSM-5	High conversion rates
Selectoforming	Ni erionite, Ni erionite/ clinoptilolite	Increase in octane number via LPG production.
Hydroisomerization	Pt mordenite	Converts low octane, pentane and hexane feeds to higher octane yields
Dewaxing	Pt mordenite, ZSM-5	Improved pour points
Benzene alkylation	ZSM-5	Ethylbenzene and styrene production with low by-product yield
Xylene isomerization	ZSM-5	Increase in *p*-xylene yield with low by-product yield
Methanol to gasoline conversion	ZSM-5	High gasoline yield with high octane rating
NO_x reduction	H mordenite	Effluent clean-up in nitric acid and nuclear reprocessing plants

Catalytic cracking

As mentioned earlier most of the world's petrol production uses zeolites to crack crude oils (i.e. to breakdown the long-chain hydrocarbons present in the oil) to useful C_1–C_6 fractions. The cracking process is increasingly being carried out in a fluidized bed of catalyst whereby preheated crude oil (370 °C) meets a catalyst bed in a 'riser'. The resulting slurry then passes into a reaction zone

(450–520 °C) and the resulting products pass into a further zone where the catalyst is separated from the products. The catalyst is steam-treated to recover sorbed hydrocarbons prior to its regeneration and reuse as a hot catalyst to preheat the crude oil entering the reaction zone. The products pass on to a fractionator from which the major products of C_1–C_3 gas, petrol, light cycle oil, heavy gas oil and recycle oil are obtained. The process is shown schematically in Fig. 91.

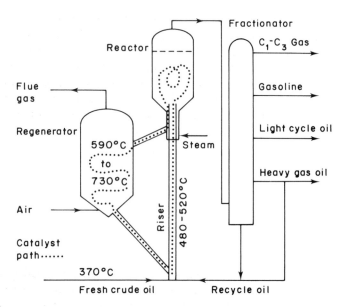

Fig. 91. Conventional layout of a fluidized bed cracking reactor and its associated processes. Reproduced by permission of the Royal Society of Chemistry from D. E. W. Vaughan in R. P. Townsend (ed.), *Properties and Applications of Zeolites*. Chemical Society Special Publication 33, 1980, p. 301

The regeneration stage is designed at 590–730 °C as an air purge to remove coke from the catalyst. This coke content is critical to the efficient running of the cracker and the reactor is designed to keep the coke level on the zeolite to below 0.1% wt carbon.

It can be seen that this regime is a harsh one and that thermal and hydrothermal stabilities of the catalyst are at a premium. Currently the preferred fluid cracking catalysts (FCC) are all faujasites—largely rare earth exchanged Y (REY) and ultrastable Y (US-Y)—but others listed in Table XXIX are commercially available and therefore presumably used.

A current trend in zeolite cracking catalyst use is the steady increase in the percentage of zeolite present in the FCC catalyst. In the early days of zeolite promoters a 5% incorporation was used, but as refinery engineers have improved processing to cope with the extreme reactivity of the zeolite in the catalyst the percentage incorporation has steadily increased to 40–50% and is expected to reach even higher levels in future 'cat' crackers.

Clearly the aim of the refinery is to maximize the yield of gasoline range products, especially those with high octane ratings (high olefine and aromatic content), and it is this area that the continuous evaluation of new zeolite catalysts is taking place. This is illustrated in Table XXIX, which emphasizes the potential of other zeolites, either alone or in admixture to improve cracking selectivities and efficiencies.

An interesting aspect of the variants possible is the ability of offretite and gmelinite to swing selectivity to the liquid propane gas (LPG) and/or petrochemical feedstock range. This property will be of importance should a future economy be less 'gasoline based' i.e. directed away from maximizing petrol production from crudes towards the formation of these other useful molecules.

Further examples of the potential for flexibility in the use of zeolites in cracking can be appreciated in the addition of small concentrations (< 10 ppm) of noble metals to promote the following reactions:

$$C + \tfrac{1}{2}O_2 \rightarrow CO$$

and

$$CO + \tfrac{1}{2}O_2 \rightarrow CO_2$$

and so aid removal of coke during the regeneration stage.

Hydrocracking

The process of hydrocracking is that in which heavy, and residual, gas oils are upgraded to central heating oil, jet and diesel fuels and petrochemical feedstocks. Catalysts containing zeolites can cope well with these lower grade oils despite the high sulphur, nitrogen and transition metal contents of the oils. The demand for this process is increasing and can be linked to the need for high yields of benzene, toluene and xylene (BTX) required for lead-free petrol and as intermediates in the production of nylon, styrene and polyesters, etc.

The process takes place on a fixed-bed catalyst operating at 250–430 °C and a hydrogen pressure of 200–2000 psi. The catalysts used are the bifunctional or dual-function ones mentioned earlier, i.e. those with hydrogenation–dehydrogenation and acidity functions.

The acidity may be zeolitic or non-zeolitic (silica/alumina or silica/zirconia) and the whole makes for a versatile catalyst composition. It enables the design of a series of catalysts of variable 'acid strength' to effect changes in selectivity. Within the zeolite promoters used, the acid functionality increasingly has its source in the use of high-silica zeolites (e.g. US-Y or dealuminated Y) which are replacing the earlier choices of CaMgY and HY. The hydrogenation–dehydrogenation properties arise from a mixed metal composition dispersed on synthetic faujasites, mordenite and erionite. Popular metal combinations are Co–Mo, Ni–W and Pt–Pd. In future process development this requirement may be fulfilled by using the hydrogen form of ZSM-5 (H-ZSM-5).

Selectoforming

Selectoforming is an industrial use of shape selectivity. It involves the selective uptake of C_5–C_9 n-paraffins from reformer product streams. Reforming is the stage in refining when naphthas from the crudes are hydrogen treated. The selectoforming catalyst selectively cracks these n-paraffins to liquid propane gas (C_1–C_3) which can be easily, separated so eliminating low octane rating fractions from the product fuel. The high octane components (branched paraffins and aromatics) are not sorbed by the zeolite and so are unchanged by the process. In this way selectoforming increases the octane number. This process is the only one in the Western world using natural zeolites—erionite or an erionite/clinoptilolite mixture containing Ni.

Hydroisomerization

This process converts hexane and pentane feeds of low octane ratings to products containing i-pentane and dimethylbutanes which have higher octane numbers. The process was developed by Shell and is known as the 'Hysomer' process, when linked to the Union Carbide 'Isosiv' i/n-paraffin separation process which uses CaA. Hydroisomerization is promoted by using a reduced Pt 'large port' mordenite which exhibits a high preference for i-pentane formation.

Dewaxing

Heavier oils, such as lubricating oils and diesel oils, contain long-chain paraffins which crystallize out below 100 °C. This affects their viscosity deleteriously as quantified by pour point determinations.

Traditionally the C_{18}, and above, hydrocarbons which give rise to pour point problems have been removed by crystallization, solvent extraction or urea adduct formation. More recently the long-chain hydrocarbons have been removed by selective hydrocracking (hydrodewaxing) with a zeolite catalysts. The basic process resembles selectoforming in that the paraffins are removed by selective sorption from the larger naphthalene and other polyaromatic molecules in the oil and then cracked to LPG in the presence of hydrogen. Originally large port mordenites, of high Si:Al ratio (40–120), doped with Pt and reduced, were preferred but more recently small crystals of ZSM-5 ($\sim 0.02\mu$) have been used. The hydrogen form of ZSM-5 achieves the catalytic dewaxing without the need to incorporate a noble metal which, of course, is a considerable advantage.

Other zeolites suggested to promote the hydrodewaxing are ferrierite, offretite, L and mazzite (Omega).

Benzene alkylation

The production of ethylbenzene from ethylene and benzene is a vital stage towards the formation of styrene for polystyrene manufacture. It takes place via the Friedel–Craft reaction catalysed by aluminium chloride:

and as such it is corrosive and produces product contamination. The Mobil 'Badger' process uses a ZSM-5-based catalyst to give a cleaner product without the problems of corrosion. Three plants, commissioned in 1978, offer a combined output of 2×10^9 lb/year towards the US capacity of about 10^{10} lb required annually.

Another useful alkylation catalysed by zeolites is the treatment of benzene with propylene to give cumene on the way to phenol and acetone production.

Xylene isomerization

Another reaction traditionally prompted by aluminium chloride (or BF_3–HF) is that of the isomerization of o-xylene to p-xylene. The o-xylene arises from the Y- or X-based separation of the xylenes described in an earlier chapter. The xylene feed for the separation comes from the cracking process.

The o-xylene is used to manufacture phthalic acid for a plasticizer intermediate, but a larger demand exists for p-xylene as a precursor to the terephthalic acid used in polyester production. Originally REY was developed to promote the *ortho/para* isomerization, but ZSM-5 is significantly better for the purpose. Here again shape selectivity can be created, with Sb_2O_3 being used to modify the diffusional properties inside the zeolite. Again it should be recalled that p-xylene has a favoured diffusion rate in an MFI structure, which means that p-xylene can be produced in high yield from large crystals of ZSM-5. In these the relative ease of escape of p-xylene from the framework shifts the equilibrium strongly to favour the product, in comparison to those which have more difficulty in diffusing out of the ZSM-5 crystals.

Methanol to gasoline conversion

During the First World War effective blockading created a critical fuel shortage in the German economy. They solved this by developing the Fischer–Tropsch process which produces hydrocarbon fuels from a feedstock of a CO and H_2 mixture. This mixture is known as 'syngas' (synthesis gas) and the CO component can be produced from any carbon source—coal in the case of Germany in 1914–18.

The main product from the Fischer–Tropsch reaction is methanol and today this can be efficiently converted to high-octane fuels by using a ZSM-5 catalyst. The process is known as the MTG (methanol to gasoline) process and uses both fixed and fluidized beds. The product is a high-quality octane fuel, with 12–14% LPG and only 1–2% methane/ethane.

132

In South Africa syngas is also produced from coal, by the SASOL process, and ZSM-5 is used in subsequent processing to improve gasoline yield and for hydrodewaxing. The SASOL process operates on a giant scale and in 1985 consumed over 30 million tons of coal. It is only economic in the context of South Africa's special needs and is not competitive with conventional fuel production via crude oil cracking. This may well change in the future and most of the developed nations have initiated MTG studies.

The New Zealand government has joined with the Mobil Company to develop a full-scale MTG plant at Motanui in North Taranaki based upon the use of ZSM-5 (note: ZSM stands for Zeolite Socony Mobil). The carbon source in this case exploits the Maui offshore natural gas field and currently produces enough petrol (1665 tons/day) to satisfy about 30% of the New Zealand demand. The syngas is produced by steam reforming the natural gas, i.e.

$$CH_4 + H_2O \rightarrow CO + H_2$$

Brazil, amongst other countries, has recognized that biomass can be the carbon source and pilot plants have been constructed on this basis. Pilot plants in the UK and elsewhere have been designed to use plastic and polymer waste as feedstock for hydrocarbon fuel production. In this and related work, such as on the catalytic breakdown of natural oils, ZSM-5 is the catalyst used.

The existence of MTG plant raises another far-reaching question. As it can be seen the zeolite catalyst used has the ability to convert oxygenated feeds to hydrocarbons. This is illustrated in Fig. 92 which also shows that by clever use of the time on stream other useful products can be collected. This offers the opportunity to use the basic process to produce chemical feedstocks rather than fuels. Of course the same choice exists in conventional oil refining and indeed the examples of alkylation and xylene isomerization constitute examples of this.

Both prospects, i.e. chemicals production from syngas and crudes, are economic options for the future and currently receive extensive study. These

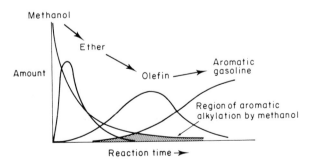

Fig. 92.   Conversion of methanol to olefins and aromatics with the alkylation process as a function of time (ZSM-5-type catalyst). Reproduced by permission of the Royal Society of Chemistry from D. E. W. Vaughan in R. P. Townsend (ed.), *Properties and Applications of Zeolites*. Chemical Society Special Publication 33, 1980, p. 318

have reached pilot plant stage and a recent plant built and operated by African Explosives and Chemical Industries (South Africa) produces 30% ethene, 20% propene and 13% LPG from an MTG process with the remaining product being petrol. Once again ZSM-5 is the catalyst. Fig. 93 shows a plan of MTC (methanol to chemicals) processes which can be expected to grow in importance, with ZSM-5-style catalysts playing an important, if not unique. part in their expansion.

   In closing this section it should be stated that in these variants of the MTG processing, no exact information on the catalysts used is forthcoming, but generally it can be expected that they will be a ZSM-5 of $Si:Al > 30$, often as H-ZSM-5, and often doped with catalytically active metals (usually noble metals).

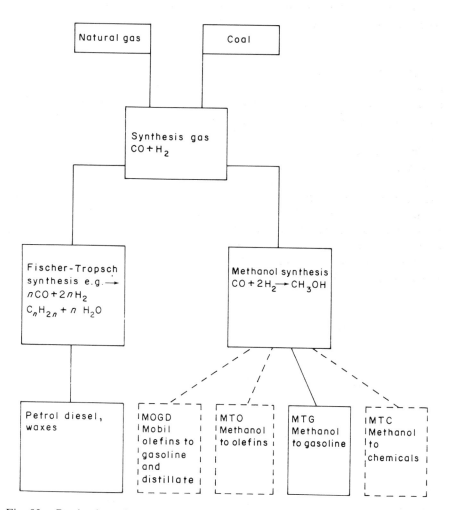

Fig. 93.   Production of synthesis gas and its routes to useful products (dashed lines signify potential processes)

NO_x reduction

In the presence of ammonia the hydrogen form of mordenite is widely used to reduce the oxides of nitrogen (NO_x) and so minimize atmospheric pollution. The prime users of this technology are nitric acid and nuclear reprocessing plants and in the reaction ammonia reacts stoichiometrically at 250 °C with the oxide to produce nitrogen and water.

An equivalent SO_x emission reduction via a zeolite catalyst is known but is apparently not used in the West, although it seems that clinoptilolite may well be installed for this purpose in some Eastern European industrial plants.

CHAPTER 10

Zeolite-like materials (zeo-types) containing elements other than Si or Al in tetrahedral framework sites

Introduction

Chapter 2 drew attention to naturally occurring minerals with elements other than Si and Al presumed to be in tetrahedral environments in their framework structures. The best known examples of such minerals are viseite and keoheite, both claimed to have phosphorus-containing analcime-type frameworks.

The existence of these materials, coupled with the wide knowledge of ions able to occupy tetrahedral sites in oxide structures, offers the prospect of synthesizing a wide range of heteroatom framework structures—hopefully with open structures appropriate to novel catalytic, ion exchange and molecular sieving properties.

Likely candidate elements for such isomorphous replacement are Ga (close to Al in Group 3 of the periodic table) and Ge (close to Si in Group 4). In addition many compounds based upon linkages of $(PO_4)^{3-}$ coordination polyhedra are well known (but not of open character except the *layer* phosphates such as zirconium phosphate).

Two general approaches have been made to place elements other than Si and Al into zeolitic structures namely by (i) direct synthesis and (ii) modification of existing zeolites.

So far as modification is concerned earlier parts of the book have noted various treatments (e.g. with P, B, etc.) to improve zeolite catalyst performance. These treatments have resulted in, for instance, B being substituted into framework sites (confirmed by MASNMR). In all these modifications the number of framework sites substituted is small—although in the case of high-silica ZSM-5 the replacement of Al by, say, B can be high, as a proportion of the original Al being replaced.

Direct synthesis has expanded considerably in the last five years and has been shown to provide a route to many more structures. This can be seen from the literature which, since 1982, has described more than 50 new prospective zeotypes, some of which have been well characterized.

This chapter will concentrate on these new phases, rather than modified zeolites, as the new materials are of more general interest. Before these are examined in more detail it should be noted that many earlier reports of zeotype synthesis are unsubstantiated and can often be explained by the presence of species encapsulated in cavities in zeolite frameworks rather than in tetrahedral sites.

Zeotypes from Union Carbide laboratories

Aluminium–phosphorus–oxygen frameworks (AlPO$_4$s)

In 1982 workers in the Union Carbide research laboratories announced the discovery of a new family of framework structures based upon Al and P. They are known by the acronym 'AlPO$_4$', the '4' being required to distinguish them from 'AlPO' which is the registered trademark of a US dog food.

To date about 24 compounds of the family have been described all having the composition of x RAl$_2$O$_3 \cdot 1.0 (\pm 0.2)$ P$_2$O$_5 \cdot y$ H$_2$O where R is the amine or quaternary ammonium species used as a component of the original gel synthesis. The quantity $(x + y)$ represents the amount needed to fill the microporous void of the final framework. Most of the AlPO$_4$s have three-dimensional structures but some are layer compounds, some are dense phosphate phases and some contain P–O–H environments.

To date 13 of the new phases can be classified as molecular sieves and their structures have been determined and shown to be three-dimensional. From the compositional formula cited above it can be implied that P and Al are present in equimolar proportions. This means that their overall framework charge is neutral and this is confirmed by their lack of ion exchange properties (even after calcination to remove R).

Examples of the structures defined are those of AlPO$_4$-5, AlPO$_4$-11 and AlPO$_4$-17 shown in Fig. 94. These are novel; other AlPO$_4$s have frameworks which can be closely related to known zeolite topologies. Table XXX summarizes the current state of knowledge of the structure and character of AlPO$_4$s.

Synthesis of AlPO$_4$s is carried out from reactive gels containing templates such as the tetrapropylammonium species. The aluminium source is critical, as is the order of addition of the reactants. Typically pseudo-boehmite (a hydrated alumina) is mixed in water with phosphoric acid to give an aluminophosphate gel to which a 'templating' molecule (R) is added. The mixture so created is allowed to crystallize in the range 125–200°C. Table XXXI gives examples of the templates used. In fact AlPO$_4$-5 has been made from many templates but other AlPO$_4$s are claimed to demonstrate clear evidence of the steric and electronic influences on the structure-directing properties of their templates.

Undoubtedly other AlPO$_4$ phases will be reported and recently workers at Virginia Polytechnic Institute (USA) have prepared a phase designated as VPI-5

Table XXX. AlPO$_4$-type microporous structures

Number	Structure type	Porosity	Pore size (nm)	Saturation H$_2$O capacity (cm^3 g^{-1})
Structure fully determined				
5	Novel	Large	0.80	0.31
11	Novel	Medium	0.60	0.16
14	Novel	Small	0.40	0.19
15	Leucophosphite[a]	—	—	—
16	Zuynite[b]	Very small	0.30	0.30
17	Erionite	Small	0.43	0.28
20	Sodalite	Very small	0.30	0.24
25	Novel	Very small	0.30	0.17
46[c]	Novel	—	—	—
Structures inferred from X-ray powder patterns				
37	Faujasite	Large	0.80	0.35
34	Chabazite	Small	0.43	0.30
35	Levynite	Small	0.43	0.30
42	[A]	Small	0.43	0.30
43	Gismondine	Small	0.43	0.34
44	Chabazite	Small	0.43	0.3–0.34
47	Chabazite	Small	0.43	0.3–0.34
Unknown structures				
36	Novel	Large	0.80	0.31
40	Novel	Large	0.70	0.33
31	Novel	Medium	0.65	0.17
41	Novel	Medium	0.60	0.22
18	Novel	Small	0.43	0.35
26	Novel	Small	0.43	0.23
33	Novel	Small	0.40	0.23
39	Novel	Small	0.40	0.23
28	Novel	Very small	0.30	0.21

[a] Leucophosphite $KFe_2[OH(H_2O)PO_4]_2 \cdot H_2O$.
[b] Zuynite $Al_{13}Si_2O_2O(OH,F)_{18}Cl$.
[c] Isomorph Co APSO-46 solved (see later).

which has been fully characterized and shown to contain a unique 18-membered oxygen ring proscribing a pore of 10.2 Å in diameter.

So far as the prospective industrial use of the AlPO$_4$ materials have been examined their thermal and hydrothermal stabilities are excellent. Structures are retained even after calcination to 1000°C and AlPO$_4$-5, -11 and -17 show no structural degradation when treated with 10% steam at 600°C. AlPO$_4$s 20, 17 and 5 have desiccant properties close to those of zeolites A and X and AlPO$_4$-11 selectively removes cyclohexane from neopentane. As might be anticipated AlPO$_4$s have only weak catalytic properties, which can be ascribed to the presence of a low concentration of surface hydroxyls (as identified by an OH band at 3680 cm^{-1}).

138

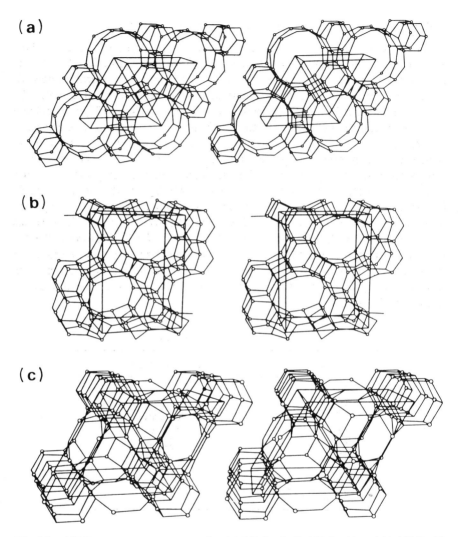

Fig. 94. AlPO$_4$ structures as stereopairs: (a) AlPO$_4$-5, (b) AlPO$_4$-11 and (c) AlPO$_4$-17. Reproduced with permission from J. M. Bennett et al., Zeolites, **6**, 351 (1986). Copyright Butterworth Publishers

Table XXXI.   Examples of AlPO$_4$ templates

Number	Template (R)
5	Tetrapropylammonium
11	Di-n-propylamine
17	Cyclohexylamine
	Quinuclidine

Silicon–aluminium–phosphorus–oxygen frameworks (SAPOs)

Shortly after the discovery of the $AlPO_4$ family the Union Carbide workers announced another group of similar compounds designated as 'SAPOs'. As the acronym implies they are structures with Si, Al and P in tetrahedral framework sites. Currently 13 microporous materials have been discovered with the composition $0.03 R(Si_x Al_y P_2)O_2$, where x varies from 0.04 to 0.20, y varies from 0.45 to 0.50 and z varies from 0.34 to 0.41. This confers a net negative charge to the framework and hence ion exchange properties. Two SAPOs have novel structures, but otherwise they are structurally related either to known zeolite topologies or to $AlPO_4$ structures. The designated numbers imply that they are of the same structural type as the corresponding $AlPO_4$. At the time of writing no full structure determination has been carried out on a SAPO and the only data published is by analogy as explained earlier and listed in Table XXXII. It has been suggested that the major replacement is that of Si for P with some of 2 Si for Al and P. Substitution of Si for P leads to Brønsted activity and SAPO-5, for instance, exhibits activity for cumene cracking and o-xylene isomerizm. The catalytic 'strength' is a function of Si concentration and structure and awaits further elucidation.

Table XXXII. SAPO structural types

Number	Structure type
40	Novel
41	Novel
34, 44	Chabazite
35	Levynite
37	Faujasite
42	A
17	Erionite
20	Sodalite
5,11,16,31	$AlPO_4$

In general the SAPOs resemble their equivalent $AlPO_4$s in sorption properties, pore size and thermal and hydrothermal stability and their synthesis follows the general style of the $AlPO_4$s.

Metal–aluminium–phosphorus–oxygen frameworks (MeAPOs)

The Union Carbide workers further extended the range of known zeotypes to another family (MeAPOs) encompassing a further 13 frameworks incorporating Co^{II}, Fe^{II}, Mn^{II}, Mg^{II}, Zn^{II} and Fe^{III}. Again some of these are related to zeolites, some are related to $AlPO_4$s and some are novel. Three have been fully elucidated, in that CoAPOs 44 and 47 are known to have chabazite topologies, with Co^{II} clearly defined in a framework site as a component of a D4R. These two cobalt

MeAPOs are a bright blue. CoAPO-50 is also blue and has a known structure which is novel and contains 12-and 8-membered 'O' rings.

The general synthetic route for the MeAPO's is like that for of the AlPO$_4$s and SAPOs and they have a compositional range of 0–0.3 R(Me$_x$ Al$_y$ P$_z$)O$_2$, where Me = Co^{II}, Fe^{II}, Mn^{II}, Mg^{II}, Zn^{II} or Fe^{III}; $x = 0.08$–0.16; $y = 0.033$–0.42 and $z \sim 0.5$. Preliminary studies show that the metal ion occupies the role of Al rather than P, so the framework carries a net negative charge, confirmed by ion exchange capacities, and high catalytic activity in some cases. Again the MeAPOs have a broad similarity to AlPO$_4$ and SAPO structural analogues so far as pore volumes are concerned, but they have much more variable thermal and hydrothermal stabilities—the Zn^{II} frameworks are of low stability for instance.

Metal–aluminium–phosphorus–silicon–oxygen frameworks (MeAPSOs)

Very recently yet another family of materials has been discovered (again at the Union Carbide laboratories) based upon the metals found in MeAPO structures but this time in Al–P *and* Si frameworks. These have been called 'MeAPSOs' and two CoAPSO structures have been fully determined—those designated 44 and 47 which have chabazite-like structures like their CoAPO analogues. MgAPSO-46 also has been structurally resolved and has a novel structure. Still further families are coded as ElAPOs and ElAPSOs and include frameworks containing Li, Be, B, Ga, As and Ti. Some contain more than one of these elements together with AlP-or Al P Si-based structures. Little has been published on the structure and chemical and physical properties of these new families but it is known that some are very highly active catalysts.

Zeotypes reported by other workers

Gallium and germanium compounds

As early as 1952 Goldsmith reported a Ga thomsonite containing Ga and since then several substitutions of Ga and Ge have been reported. These are summarized in Table XXXIII. The general syntheses are of zeolitic type but are often not 'templated'. Included in the products is one pure Ge analogue of

Table XXXIII. Gallium and germanium zeotypes

Ga-containing	Ge-containing	Ga and Ge-containing
Thomsonite	Harmotome	Faujasite
Faujasite	Faujasite	A
A	A	
Harmotome	Phillipsite	
Sodalite	ZSM-5	
ZSM-5		

faujasite—$Na_8[Ge_6Si_6O_{24}]\cdot 6H_2O$—which has been studied as a catalyst and molecular sieve. Ga sodalite has been fully elucidated by X-ray analysis. A gallium phosphate isomorph of $AlPO_4$-14 has been used to define the structure of $AlPO_4$-14.

Beryllium and boron compounds

Both analcime and faujasite structures containing Be have been reported and boron analogues of analcime and Na-P claimed. These reports have not been confirmed, as yet, by full structure determination and their properties remain largely unstudied.

ZSM-5 structures containing only B and Si frameworks, described as 'boralites' have been synthesized and characterized. Their synthesis uses a template and the observed increase in unit cell dimensions with increased B content has been presented (with MASNMR data) as confirmation of B as a framework atom. An Fe^{III} analogue of analcime is known, i.e. $KFeSi_2O_6$, and a ferrisilicate of ZSM-5 structure has been well characterized. Claims to have included species such as Cr^{III} and Ti^{III} into ZSM-5 tetrahedral environments need more evidence.

Other workers report zinco- and stanno-silicates with zeolitic properties but have not confirmed that Zn and Sn are in tetrahedral sites.

No doubt the near future will produce many more novel zeotypes and this represents an exciting new range of materials with potentially useful applications as catalysts, molecular sieves and perhaps ion exchangers.

Selected bibliography*

Books

1. D. W. Breck, *Zeolite Molecular Sieves*, Wiley, New York, 1974.
2. P. A. Jacobs, *Carboniogenic Activity of Zeolites*, Elsevier, Amsterdam, 1977.
3. R. M. Barrer, *Zeolites and Clay Minerals as Sorbents and Molecular Sieves*, Academic Press, London, 1978.
4. R. M. Barrer, *Hydrothermal Chemistry of Zeolites*, Academic Press, London, 1982.
5. D. M. Ruthven, *Principles of Adsorption and Adsorption Processes*, Wiley, New York, 1985.
6. G. Gottardi and E. Galli, *Natural Zeolites*, Springer Verlag, Berlin, 1985.

Conference proceedings

1. †*Molecular Sieves* (London Conf.), Society of Chemistry and Industry, 1968.
2. †*Molecular Sieves 1 & 2* (Worcester (USA) Conf.), ACS Advances in Chemistry Series, Vols. 101 and 102, 1971.
3. †*Molecular Sieves* (Zurich Conf.) (Eds. W. M. Meier and J. B. Uytterhoeven), ACS Advances in Chemistry Series, Vol. 121, 1973.
4. †*Molecular Sieves*—11 (Chicago Conf.) (Ed. J. R. Katzer) ACS Symposium Series, Vol. 40, 1977.
5. †*Proc. 5th Int. Zeolite Conf.* (Naples Conf.) (Ed. L. V. C. Rees), Heyden, London, 1980.
6. †*Proc. 6th Int. Zeolite Conf.* (Reno Conf.) (Eds. A. Bisio and D. H. Olson), Butterworth Scientific Ltd., Guildford, 1984.
7. †*New Developments in Zeolite Science and Technology* (Tokyo Conf.), (Eds. Y. Murakami, A. Iijama and J. W. Ward), Elsevier, Amsterdam and Kodansha, Tokyo, 1986.
8. *Natural Zeolites: Occurrence, Properties and Use* (Eds. L. B. Sand and F. A. Mumpton), Pergamon Press, Oxford, 1978.
9. *Properties and Applications of Zeolites* (Ed. R. P. Townsend), The Chemical Society, London, 1980.
10. *Metal Microstructures in Zeolites* (Eds. P. A. Jacobs, N. I. Jaeger, P. Jiru and G. Schultz-Ekloff), Elsevier, Amsterdam, 1982.
11. *Intrazeolite Chemistry* (Eds. G. D. Stucky and F. G. Dwyer), ACS Symposium Series, Vol. 218, 1983.
12. *Structure and Reactivity of Modified Zeolites* (Eds. P. A. Jacobs, N. I. Jaeger, P. Jiru, V. B. Kazansky and G. Schultz-Ekloff), Elsevier, Amsterdam, 1984.

---

* A complete bibliography of all publications concerned with zeolites has been compiled by A. E. Comyns and is published (Part I Publications in English) in *Zeolites*, **5**, 342(1985) and (Parts II-IV all books (including those in languages other than English) in *Zeolites*, **8**, 255 (1988).

† Conferences organized by the International Zeolite Association.

13. *Zeo-Agriculture: Use of Natural Zeolites in Agriculture and Aquaculture* (Eds. W. G. Pond and F. A. Mumpton), Westview Press, Boulder, Colorado, 1984.

Others

1. J. A. Rabo (Ed.), *Zeolite Chemistry and Catalysis*, ACS Monograph No 171, 1976.
2. F. A. Mumpton (Ed.), *Mineralogy and Geology of Natural Zeolites*, Mineralogical Society of America, Short Course Notes, Vol. 4, 1977.
3. C. W. Roberts, in *Speciality Chemicals Protection, Marketing and Applications*, Chemical Society, London, 1981.
4. W. J. Mortier, *Compilation of Extra Framework Sites in Zeolites*, Butterworth Scientific Ltd., Guildford, 1982.
5. R. Von Ballmoos, *Collection of Simulated XRD Patterns for Zeolites*, Butterworth Scientific Ltd., Guildford, 1984.
6. *Chem. Ind.*, *No.* 7 (1984) (5 *review articles*).
7. Vedrine J. C., in *Solid State in Catalysis* (Eds. R. K. Grasselli and J. F. Brazdil) ACS Symposium Series, Vol. 279, 1985.
8. W. M. Meier and D. J. Olson, *Atlas of Zeolite Structure Types* Butterworth Scientific Ltd., Sevenoaks, 1988.

Index

145